Dry Farming

for Sustainable Agriculture

Dry Farming

For Sustainable Agriculture

John A. Widtsoe

AGROBIOS (INDIA)

Published by:
AGROBIOS (INDIA)
Agro House, Behind Nasrani Cinema
Chopasani Road, Jodhpur 342 002
Phone: 91-0291-2642319, Fax: 2643993
E. mail: agrobios@sify.com

AGROBIOS (INDIA)

First Indian Impression: 2005
Reprint: 2007
Reprint: 2009

***Price:* Rs. 795.00 / US$ 40.00**

Published by: Dr. Updesh Purohit for Agrobios (India), Jodhpur
Lasertypeset at: Yashee Computers and Printers, Jodhpur
Cover Design by: Reena
Printed at: Bharat Printing Press, Jodhpur

PREFACE

Nearly Six tenths of the earth's land surface receive an annual rainfall of less than twenty inches, and can be reclaimed for agricultural purposes only by irrigation and dry-farming. A perfected world-system of irrigation will convert about one tenth of this vast area into an incomparably fruitful garden, leaving about one half of the earth's land surface to be reclaimed, if at all, by the methods of dry-farming. The noble system of modern agriculture has been constructed almost wholly in countries of abundant rainfall, and its applications are those demanded for the agricultural development of humid regions. Until recently, irrigation was given scant attention, and dry-farming, with its world problem of conquering one half of the earth, was not considered. These facts furnish the apology for the writing of this book.

One volume, only, in this world of many books, and that less than a year old, is devoted to the exposition of the accepted dry-farm practices of to-day.

The book now offered is the first attempt to assemble and organize the known facts of science in their relation to the profitable production of plants, without irrigation, in regions of limited rainfall. The needs of the actual farmer, who must understand the principles before his practices can be wholly satisfactory, have been kept in view primarily; but it is hoped that the enlarging group of dry-farm investigators will also be helped by this presentation of the principles of dry-farming. The subject is now growing so rapidly that there will soon be room for two classes of treatment: one for the farmer, and one for the technical student.

This book has been written far from large libraries, and the material has been drawn from the available sources. Specific references are not given in the text, but the names of investigators or institutions are found with nearly all statements of fact. The files of the Experiment Station Record and Der Jahresbericht der Agrikultur Chemie have taken the place of the more desirable original publications. Free use has been made of the publications of the experiment stations and the United States Department of Agriculture. Inspiration and suggestions have been sought and found constantly in the works of the princes of American soil investigation, Hilgard of California and King of Wisconsin. I am under deep obligation, for assistance rendered to numerous friends in all parts of the country, especially to Professor L. A. Merrill, with whom I have collaborated for many years in the study of the possibilities of dry-farming in Western America.

The possibilities of dry-farming are stupendous. In the strength of youth we may have felt envious of the great ones of old; of Columbus looking upon the shadow of the greatest continent; of Balboa shouting greetings to the resting Pacific; of Father Escalante, pondering upon the mystery of the world, alone, near the shores of America's Dead Sea. We need harbor no such envyings, for in the conquest of the nonirrigated and nonirrigable desert are offered as fine opportunities as the world has known to the makers and shakers of empires. We stand before an undiscovered land; through the restless, ascending currents of heated desert air the vision comes and goes. With striving eyes the desert is seen covered with blossoming fields, with churches and homes and schools, and in the distance, with the vision is heard the laughter of happy children.

The desert will be conquered.

JOHN A. WIDTSOE

CONTENTS

CHAPTER - I

DRY-FARMING: INTRODUCTION

DRY-FARMING DEFINED

Dry-Farming, as at present understood, is the profitable production of useful crops, without irrigation, on lands that receive annually a rainfall of 20 inches or less. In districts of torrential rains, high winds, unfavorable distribution of the rainfall, or other water-dissipating factors, the term "dry-farming" is also properly applied to farming without irrigation under an annual precipitation of 25 or even 30 inches. There is no sharp demarcation between dry and humid-farming.

When the annual precipitation is under 20 inches, the methods of dry-farming are usually indispensable. When it is over 30 inches, the methods of humid-farming are employed; in places where the annual precipitation is between 20 and 30 inches, the methods to be used depend chiefly on local conditions affecting the conservation of soil moisture. Dry-farming, however, always implies farming under a comparatively small annual rainfall.

The term "dry-farming" is, of course, a misnomer. In reality it is farming under drier conditions than those prevailing in the countries in which scientific agriculture originated. Many suggestions for a better name have been made. "Scientific agriculture" has been proposed, but all agriculture should be scientific, and agriculture without irrigation in an arid country ahs no right to lay sole claim to so general a title. "Dry-land agriculture", which has also been suggested, is no improvement over "Dry-farming" as it is longer and also carries with it the idea of dryness. Instead of the name "dry-farming" it would, perhaps, be better to use the names, "arid-farming" "semiarid-farming", "humid-farming", and "irrigation-farming", according to the climatic conditions prevailing in various parts of the world. However, at the present time the name "dry-farming" is in such general use that it would seem unwise to suggest any change. It should be used with the distinct understanding that as far as the word "dry" is concerned it is a misnomer. When the two words are hyphenated, however, a compound technical term- "dry-farming" – is secured which has a meaning of its own, such as we have just defined it to be; and "dry-farming", therefore, becomes an addition to the lexicon.

Fig. 1: Breaking the typical prairie land of the Rocky Mountain foothills. Montana, 1908.

Dry-versus Humid-farming

Dry-farming, as a distinct branch of agriculture, has for its purpose the reclamation, for the use of man, of the vast un-irrigable "desert" or "semi-desert" areas of the world, which until recently were considered hopelessly barren. The great underlying principles of agriculture are the same the world over, yet the emphasis to be placed on the different agricultural theories and practices must be shifted in accordance with regional conditions. The agricultural problem of first importance in humid regions is the maintenance of soil fertility; and since modern agriculture was developed almost wholly under humid conditions, the system of scientific agriculture has for its central idea the maintenance of soil fertility. In arid regions, on the other hand, the conservation of the natural water precipitation for crop production is the important problem; and a new system of agriculture must therefore be constructed, on the basis of the old principles, but with the conservation of the natural precipitation as the central idea. The system of dry-farming must marshal and organize all the established facts of science for the better utilization, in plant growth, of a limited rainfall. The excellent teachings of humid agriculture respecting the maintenance of soil fertility will be of high value in the development of dry-farming, and the firm establishment of right methods of conserving and using the natural precipitation will undoubtedly have a beneficial effect upon the practice of humid agriculture. Figures 1-2 show some of the characteristic features of dry-farming regions.

The problems of dry-farming

The dry-farmer, at the outset, should know with comparative accuracy the annual rainfall over the area that he intends to cultivate. He must also have a good acquaintance with the nature of the soil, not only as regards its plant-food content, but as to its power to receive and retain the water from rain and snow. In fact, a knowledge of the soil is indispensable in successful dry-farming. Only by such knowledge of the rainfall and the soil is he able to adapt the principles outlined in this volume to his special needs.

Since, under dry-farm conditions, water is the limiting factor of production, the primary problem of dry-farming is the most effective storage in the soil of the natural precipitation. Only the water, safely stored in the soil within reach of the roots, can be used in crop production. Of nearly equal importance is the problem of keeping the water in the soil until it is needed by plants. During the growing season, water may be lost from the soil by downward drainage or by evaporation from the surface. It becomes necessary, therefore, to determine under what conditions the natural precipitation stored in the soil moves downward and by what means surface evaporation may be prevented or regulated. The soil-water, of real use toe plants, is that taken up by the roots and finally evaporated from the leaves. A large part of the water stored in the soil is thus used. The methods whereby this direct draft of plants on the soil-moisture may be regulated are, naturally, of the utmost importance to the dry-farmer, and they constitute another vital problem of the science of dry-farming.

The relation of crops to the prevailing conditions of arid lands offers another group of important dry-farm problems. Some plants use much less water then others. Some attain maturity quickly, and in that way become desirable for dry-farming. Still other crops, grown under humid conditions, may easily be adapted to dry-farming conditions, if the correct methods are employed, and in a few seasons maybe made valuable dry-farm crops. The individual characteristics of each crop should be known as they relate themselves to a low rainfall and arid soils.

After a crop has been chosen, skill and knowledge are needed in the proper seeding, tillage, and harvesting of the crop. Failures frequently result from the want of adapting the crop treatment to arid conditions.

After the crop has been gathered and stored, its proper use is another problem for the dry-farmer. The composition of dry-farm crops is different from that of crops grown with an abundance of water. Usually, dry-farm crops are much more nutritious and therefore should command a higher price in the markets, or should be fed to stock in corresponding proportions and combinations.

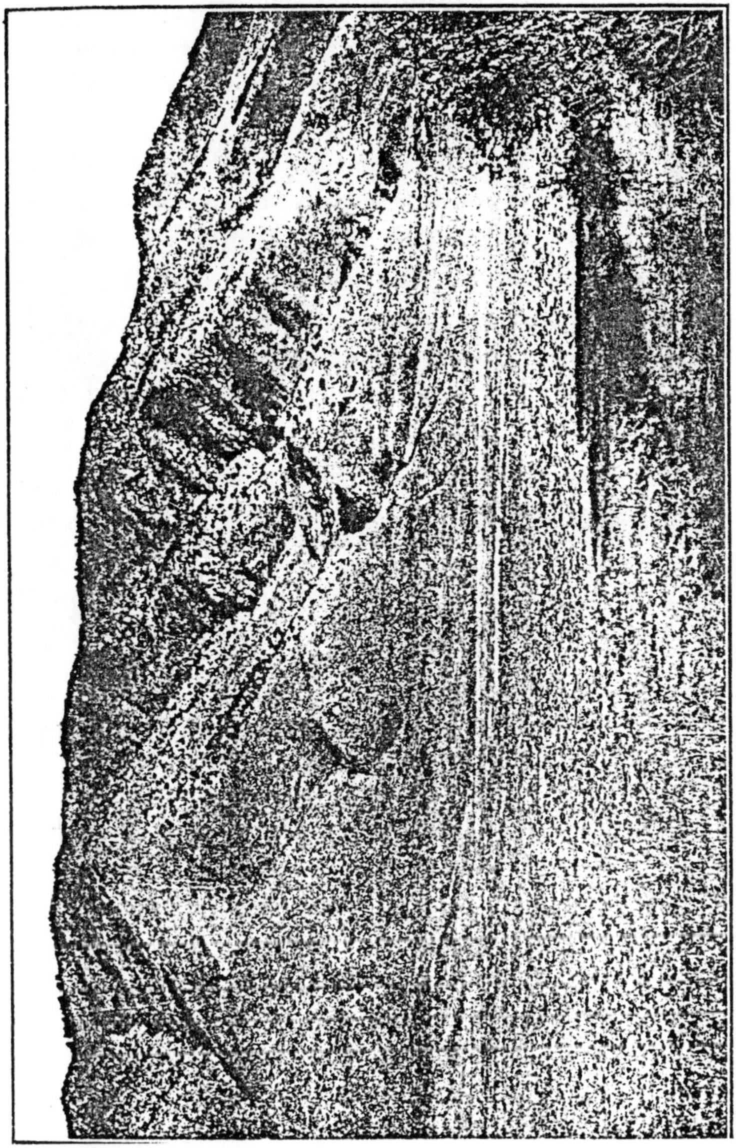

Fig. 2: Elevated lands. Much of the best dry-farm land lies above the canal. Note the canal in the lower right-hand corner.

The fundamental problems of dry-farming are, then the storage in the soil of a small annual rainfall; the retention in the soil of the moisture until it is needed by plants; the prevention of the direct evaporation of soil-moisture during the growing season; the regulation of the amount of water drawn from the soil by plants; the choice of crops suitable for growth under arid conditions; the application of suitable crop treatments, and the disposal of dry-farm products, based upon the superior composition of plants grown with small amounts of water. Around these fundamental problems cluster a host of minor, though also important, problems. When the methods of dry-farming are understood and practiced, the practice is always successful; but it requires more intelligence, more implicit obedience to nature's laws, and greater vigilance, than farming in countries of abundant rainfall.

The chapters that follow will deal almost wholly with the problems above outlined as they present themselves in the construction of a rational system of farming without irrigation in countries of limited rainfall.

CHAPTER II

THE THEORETICAL BASIS OF DRY-FARMING

The confidence with which scientific investigators, familiar with the arid regions, have attacked the problems of dry-farming resists largely on the known relationship of the water requirements of plants to the natural precipitation of rain and snow. It is a most elementary fact of plant physiology that no plant live and grow unless it has at its disposal a sufficient amount of water.

The water used by plants is almost entirely taken from the soil by the minute root-hairs radiating from the roots. The water thus taken into the plants is passed upwards through the stem to the leaves, where it is finally evaporated. There is, therefore, a more or less constant stream of water passing through the plant from the roots to the leaves.

By various methods it is possible to measure the water thus taken from the soil. While this process of taking water from the soil is going on within the plant, a certain amount of soil-moisture is also lost by direct evaporation from the soil surface. In dry-farm sections, soil-moisture is lost only by these two methods; for wherever the rainfall is sufficient to cause drainage from deep soils, humid conditions prevail.

Water for one pound dry matter

Many experiments have been conducted to determine the amount of water of water used in the production of one pound of dry plant substance. Generally, the method of experiments has been to grow plants in large pots containing weighted quantities of soil. As needed, weighed amounts of water were added to the pots. To determine the loss of water, the pots were weighed at regular intervals of three days to one week. At harvest time, the weight of dry matter was carefully determined for each pot. Since the water lost by the pots was also known, the pounds of water used for the production of every pound of dry matter were readily calculated.

The first reliable experiments of the kind were undertaken under humid conditions in Germany and other European countries. From the mass of results, some have been selected and presented in the following table. The work was done by the famous German investigators, Wollny, Hellreigel, and Sorauer, in the early eighties of the last century. In every case, the numbers in the table represent the number of pounds of water used for the production of one pound of ripened dry substance:

Fig. 3: Weighing pots to determine the amount of water required for the production of plants. Utah, 1901

Pounds of water for one pound of dry matter

TABLE: 1

	Wollny	Hellriegel	Sorauer
Wheat	-	338	459
Oats	665	376	569
Barley	-	310	431
Rye	774	353	236
Corn	233	-	-
Buckwheat	646	393	-
Peas	416	273	-
Horsebeans	-	282	-
Red clover	-	310	-
Sunflowers	490	-	-
Millet	447	-	-

It is clear from the above results, obtained in Germany, that the amount of water required to produce a pound of dry matter is not the same for all plants, nor is it the same under all conditions for the same plant. In fact, as will be shown in a later chapter, the water requirements of any crop depend upon numerous factors, more or less controllable. The range of the above German results is from 233 to 774 pounds, with an average of about 419 pounds of water for each pound of dry matter produced.

During the late eighties and early nineties, King conducted experiments similar to the earlier German experiments, to determine the water requirements of crops under Wisconsin conditions. A summary of the result of these extensive and carefully conducted experiments is as follows:

Oats	385
Barley	464
Corn	271
Peas	477
Clover	576
Potatoes	385

The figures in the above table, averaging about 446 pounds, indicate that very nearly the same quantity of water is required for the production of crops in Wisconsin as in Germany. The Wisconsin results tend to be somewhat higher than those obtained in Europe, but the difference is small.

Fig. 4: Plant house at Wisconsin in which F. H. King did much of his famous work on the water requirements of plants.

It is a settled principle of science, as will be more fully discussed later, that the amount of water evaporated from the soil and transpired by plant leaves increases materially with an increase in the average temperature during the growing season, and is much higher under a clear sky and in districts where the atmosphere is dry. Wherever dry-farming is likely to be practiced, a moderately high temperature, a cloudless sky, and a dry atmosphere are the prevailing conditions. The appeared probable, therefore, that in arid countries the amount of water required for the production of one pound of dry matter would be higher than in the humid regions of Germany and Wisconsin. To secure information on this subject, Widtose and Merrill undertook, in 1900, a series of experiments in Utah, which were conducted upon the plan of the earlier experimenters. An average statement of the results of six years experimentation is given in the subjoined table, showing the number of pounds of water required for one pound of dry matter on fertile soils:

Wheat	1048
Corn	589
Peas	1118
Sugar	630

These Utah findings support strongly the doctrine that the amount of water required for the production of each pound of dry matter is very much larger under arid conditions, as in Germany or Wisconsin. It must be observed, however, that in all of these experiments the plants were supplied with water in a somewhat wasteful manner; that is, they were given an abundance of water, and used the largest quantity possible under the prevailing conditions. No attempt of any kind was made to economize water. The results, therefore, represent maximum results and can be safely used as such. Moreover, the methods of dry-farming, involving the storage of water in deep soils and systematic cultivation, were not employed. The experiments, both in Europe and America, rather represent irrigated conditions. There are good reasons for believing that in Germany, Wisconsin, and Utah the amount above given can be materially reduced by the employment of proper cultural methods.

In view of these findings concerning the water requirements of crops, it cannot be far from the truth to say that, under average cultural conditions, approximately 750 pounds of water are required in an arid district for the production of one pound of dry matter (Fig. 5). Where the aridity is intense, this figure may be somewhat low, and in localities of sub-humid conditions, it will undoubtedly be too high. As a maximum average, however, for districts interested in dry-farming, it can be used with safety.

Crop-producing power of rainfall

If this conclusion, that no more than 750 pounds of water are required under ordinary dry-farm conditions for the production of one pound of dry matter, be accepted, certain interesting calculations can be made respecting the possibilities of dry-farming. For example, the production of one bushel of wheat will require 60 times 750, or 45,000 pounds of water. The wheat kernels, however, cannot be produced without a certain amount of straw, which under conditions of dry-farming seldom forms quite one half of the weight of the whole plant.

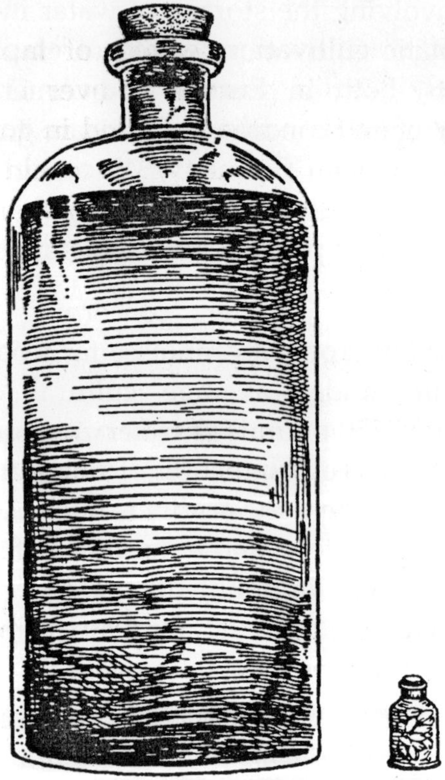

Fig. 5: The water in the large bottle would be required to produce the grain in the small bottle.

Let us say, however, that the weights of straw and kernels are equal. Then, to produce one bushel of wheat, with the corresponding quantity of straw, would require 2 times 45,000 or 90,000 pounds of water. This is equal to 45 tons of water for each bushel of wheat. While this is a large figure, yet, in many localities, it is undoubtedly well within the truth. In comparison with the amount of water that fall upon the land as rain, it does not seem extraordinarily large.

One inch of water over one acre of land weighs approximately 226,875 pounds, or over 113 tons. If this quantity of water could be stored in the soil and used wholly for plant production, it would produce, at the rate of 45 tons of water for each bushel, about 2½ bushels of what. With 10 inches of rainfall, which up to the present seems to be the lower limit of successful dry-farming, there is a maximum possibility of producing 25 bushes of wheat annually.

In the subjoined table, constructed on the basis of the discussion of this chapter, the wheat-producing powers of various degrees of annual precipitation are shown:-

1. One acre inch of water will product 2½ bushels of wheat.

2. Ten acre inches of water will produce 25 bushels of wheat.

3. Fifteen acre inches of water will produce 37½ bushels of wheat.

4. Twenty acre inches of water will produce 50 bushels of wheat.

It must be distinctly remembered, however, that under no known system of tillage can all the water that falls upon a soil be brought into the soil and stored there for plant use. Neither is it possible to treat a soil so that all the stored soil-moisture maybe used for plant production. Some moisture, of necessity, will evaporate directly from the soil, and some may be lost in many other ways. Yet, even under a rainfall of 12 inches, if only one half of the water can be conserved, which experiments have shown to be very feasible, there is a possibility of producing 30 bushels of wheat per acre every other year, which insures an excellent interest on the money and labor invested in the production of the crop.

It is one the grounds outlined in this chapter that students of the subject believe that ultimately large areas of the desert may be reclaimed by means of dry-farming. The real question before the dry-farmer is not, "Is the rainfall sufficient?" but rather, "Is it possible so to conserve and use the rainfall as to make it available for the production of profitable crops?".

CHAPTER III

DRY-FARM AREAS – RAINFALL

The annual precipitation of rain and snow determines primarily the location of dry-farm areas. As the rainfall varies, the methods of dry-farming must be varied accordingly. Rainfall, alone, does not, however, furnish a complete index of the crop-producing possibilities of a country.

The distribution of the rainfall, the amount of snow, the water-holding power of the soil, and the various moisture-dissipating causes, such as winds, high temperature, abundant sunshine, and low humidity, frequently combine to offset the benefits of a large annual precipitation. Nevertheless, no one climatic feature represents, on the average, so correctly dry-farming possibilities as does the annual rainfall. Experience has already demonstrated that wherever the annual precipitation is above 15 inches, there is no need of crop failures, if the soils are suitable and the methods of dry-farming are correctly employed.

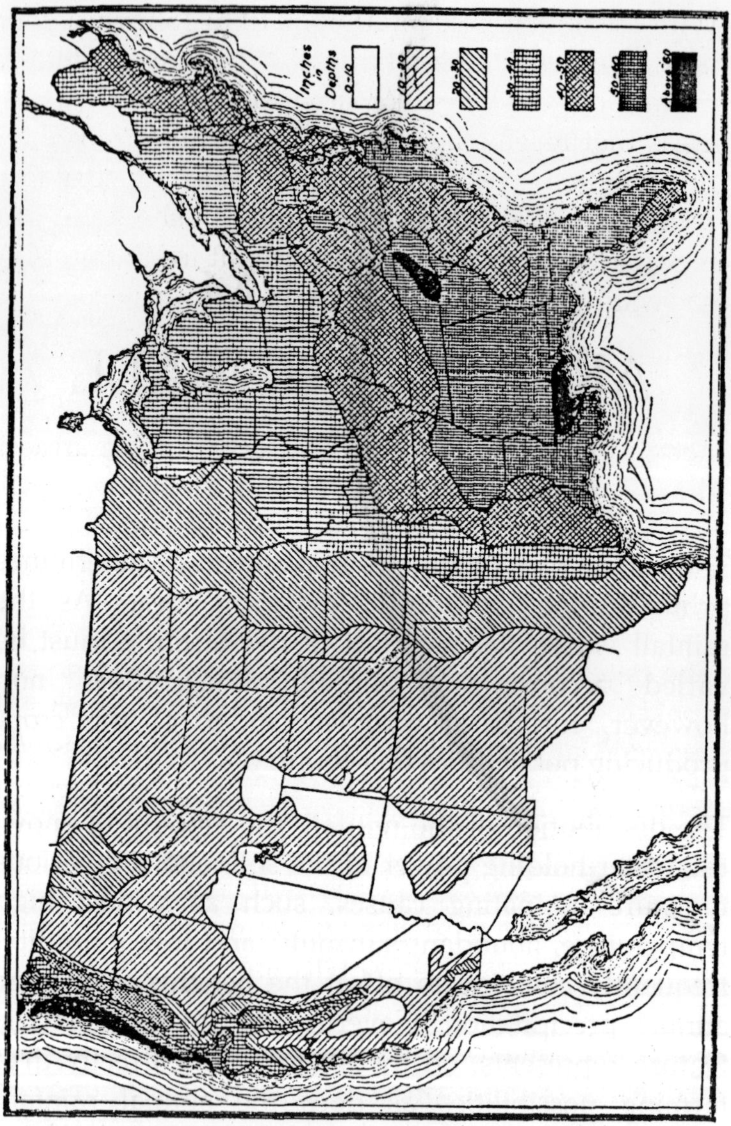

Fig. 6: The annual rainfall over the United States. (After Henry).

With an annual precipitation of 10 to 15 inches, there need be very few failures, if proper cultural precautions are taken. With our present methods, the areas that receive less than 10 inches of atmospheric precipitation per year are not safe for dry-farm purposes. What the future will show in the reclamation of these deserts, without irrigation, is yet conjectural.

Arid, semiarid, and sub-humid

Before proceeding to an examination of the areas in the United States subject to the methods of dry-farming, it may be well to define somewhat more clearly the terms ordinarily used in the description of the great territory involved in the discussion.

The states lying west of the 100th meridian are loosely spoken of as arid, semiarid, or sub-humid states. For commercial purposes no state wants to be classed as arid and to suffer under the handicap of advertised aridity. The annual rainfall of these states ranges from about 3 to over 30 inches.

In order to arrive at greater definiteness, it may be well to assign definite rainfall values to the ordinarily used descriptive terms of the region in question. It is proposed, therefore, that districts receiving less than 10 inches of atmospheric precipitation annually, be designated *arid;* those receiving between 10 and 20 inches, *semiarid;* those receiving between 20 and 30 inches, *sub-humid,* and those receiving over 30 inches, *humid.* It is admitted that even such a classification is arbitrary, since aridity does not alone depend upon the rainfall, and even under such a classification there is an unavoidable overlapping. However, no one factor so fully represents varying degrees of aridity as the annual precipitation, and there is a great need for concise definitions of the terms used in describing the parts of the country that come under dry-farming discussions. In the volume, the terms "arid", "semiarid", "sub-humid" and "humid" are used as above defined.

Precipitation over the dry-farm territory

The map (Fig. 6), based upon the work of Professor A. J. Henry of the United States Weather Bureau, shows graphically the normal annual precipitation in the United States of America. Examination of this map proves that nearly on half of the whole area of the United States receives 20 inches or less rainfall annually; and that when the strip receiving between 20 and 30 inches is added, the whole area directly subject to reclamation by irrigation or dry-farming is considerably more than one half (63 per cent) of the whole area of the United States.

Eighteen states are included in this area of low rainfall. The areas of these, as given by the Census of 1900, grouped according to the annual precipitation received, are shown in Table 2.

The territory directly interested in the development of the methods of dry-farming forms 63 per cent of the whole of the continental United States, not including Alaska, and covers an area of 1,861,652 square miles, or 1,191,457,280 acres.

TABLE: 2

	Total Area Land Surface (Sq. Miles)	
	Arid to Semi-Arid Group	
Arizona	112,920	
California	156,172	
Colorado	106,645	
Idaho	84,290	
Nevada	109,740	
Utah	82,190	
Wyoming	97,575	
Total		**746,532**
	Semiarid to Sub-humid Group	
Montana	145,3101	
Nebraska	76,840	
New Mexico	112,460	
North Dakota	70,195	
Oregon	94,560	
South Dakota	76,850	
Washington	66,880	
Total		**653,095**
	Sub-Humid to Humid Group	
Kansas	81,700	
Minnesota	79,205	
Oklahoma	38,830	
Texas	262,290	
Total		**462,025**
Total for all groups		**1,861,652**

If any excuse were needed for the lively interest taken in the subject of dry-farming, it is amply furnished by these figures showing the vast extent of the country interested in the reclamation of land by the methods of dry-farming. As will be shown below, nearly every other large country possesses similar immense areas under limited rainfall.

Of the one billion, one hundred and ninety-one million, four hundred and fifty-seven thousand, two hundred and eighty acres (1,191,457,280) representing the dry-farm territory of the United States, about 22 per cent, or a little more than one fifth, is sub-humid and receives between 20 and 30 inches of rainfall, annually; 61 per cent, or a little more than three fifths, is semiarid and receives between 10 and 20 inches, annually, and about 17 per cent, or a little less than one fifth, is arid and receives less than 10 inches of rainfall, annually.

These calculations are based upon the published average rainfall maps of the United States Weather Bureau. In the far West, and especially over the so-called "desert" regions, with their sparse population, meteorological stations are not numerous, nor is it easy to secure accurate data from them. It is strongly probable that as more stations are established, it will be found that the area receiving less than 10 inches of rainfall annually is considerably smaller than above estimated. In fact, the United States Reclamation Service states that there are only 70,000,000 acres of desert-like land; that is, land which does not naturally support plants suitable for forage. This area is about one third of the lands which, so far as known, at present receive less than 10 inches of rainfall, or only about 6 per cent of the total dry-farming territory.

In any case, the semiarid area is at present most vitally interested in dry-farming. The sub-humid area need seldom suffer from drought, if ordinary well-known methods are employed; the arid areas, receiving less than 10 inches of rainfall, in all probability, can be reclaimed without irrigation only by the development of more suitable methods than are known to-day. The semiarid area, which is the special consideration of present-day dry-farming, represents an area of over 725,000,000 acres of land. Moreover, it must be remarked that the full certainty of crops in the sub-humid regions will come only with the adoption of dry-farming methods; and that results already obtained on the edge of the "deserts" lead to the belief that a large portion of the area receiving less than 10 inches of rainfall, annually, will ultimately be reclaimed without irrigation.

Naturally, not the whole of the vast area just discussed could be bought under cultivation, even under the most favorable conditions of rainfall. A very large portion of the territory in question is mountainous and often of so rugged a nature that to farm it would be an impossibility. It must not be forgotten, however, that some of the best dry-farm lands of the West are found in the small mountain valleys, which usually are pockets of most fertile soil, under a good supply of rainfall. The foothills of the mountains are almost invariably excellent dry-farm lands. Newell estimates that 195,000,000 acres of land in the arid to sub-humid sections are covered with a more or less dense growth of timber. This timbered area roughly represents the mountainous and therefore the nonarable portions of land. The same authority estimates that the desert-like lands cover an area of 70,000,000 acres. Making the most liberal estimates for mountainous and desert-like lands, at least one half of the whole area, or about 600,000,000 acres, is arable land, which by proper methods may be reclaimed for agricultural purposes. Irrigation when fully developed may reclaim not to exceed 5 per cent of this area. From any point of view, therefore, the possibilities involved in dry-farming in the United States are immense.

Dry-farm area of the world

Dry-farming is a world problem. Aridity is a condition met and to be overcome upon every continent. McColl estimates that in Australia, which is somewhat larger than the continental United States of America, only one third of the whole surface receives above 20 inches of rainfall annually; one third receives from 10 to 20 inches, and one third receives less than 10 inches. That is, about 1,267,000,000 acres in Australia are subject to reclamation by dry-farming methods. This condition is not far from that which prevails in the United States, and is representative of every continent of the world. The map no pages 30-31 shows graphically the approximate rainfall over the various parts of the earth's land surface. The following table gives the proportions of the earth's land surface under various degrees of annual precipitations:-

TABLE: 3

Annual precipitation	Proportion of Earth's land surface
Under 10 inches	25.0 per cent
From 10 to 20 inches	30.0 per cent
From 20 to 40 inches	20.0 per cent
From 40 to 60 inches	11.0 per cent
From 60 to 80 inches	9.0 per cent
From 80 to 120 inches	4.0 per cent
From 120 to 160 inches	0.5 per cent
Above 160 inches	0.5 per cent
	100.00

Fifty-five per cent, or more, than one half of the total land surface of the earth, receives an annual precipitation of less than 20 inches, and must be reclaimed, if at all, by dry-farming. At least 10 per cent more receives from 20 to 30 inches under conditions that make dry-farming methods necessary. A total of about 65 per cent of the earth's land surface is, therefore, directly interested in dry-farming. With the future perfected development of irrigation systems and practices, not more than 10 per cent will be reclaimed by irrigation. Dry-farming is truly a problem to challenge the attention of the race.

CHAPTER IV

DRY-FARM AREAS – GENERAL CLIMATIC FEATURES

The dry-arm territory of the United States stretches from the Pacific seaboard to the 96th parallel of longitude, and from the Canadian to the Mexican boundary, making a total area of nearly 1,800,000 square miles. Shi immense territory is far from being a vast level plain. On the extreme east is the Great Plains region of the Mississippi Valley which is a comparatively uniform country of rolling hills, but no mountains. At a point about one third of the whole distance westward the whole land is lifted skyward by the Rocky Mountains, which cross the country from south to northwest. Here are innumerable peaks, canons, high table-lands, roaring torrents, and quiet mountain valleys. West of the Rockies is the great depression known as the Great Basin, which has no outlet to the ocean. It is essentially a gigantic level lake floor traversed in many directions by mountain ranges that are off-shoots from the backbone of the Rockies. South of the Great Basin are the high plateaus, into which many great chasms are cut, the best known and largest of which is the great Canon of the Colorado. North and east of the Great Basin is the Columbia River Basin characterized by basaltic rolling plains and broken mountain country. To the west, the floor of the Great Basin is lifted up into the region of eternal snow by the Sierra Nevada Mountains which north of Nevada are known as the Caseadses. On the west, the Sierra Nevadas slope gently, through intervening valleys and minor mountain ranges, into the Pacific Ocean. It would be difficult to imagine a more diversified topography than is possessed by the dry-farm territory of the United States.

Uniform climatic conditions are not to be expected over such a broken country. The chief determining factors of climate-latitude, relative distribution of land and water, elevation, prevailing winds-swing between such large extremes that of necessity the climatic conditions of different sections are widely divergent. Dry-farming is so intimately related to climate that the typical climatic variations must be pointed out.

The total annual precipitation is directly influenced by the land topography, especially by the great mountain ranges. On the east of the Rocky Mountains is the sub-humid district, which receives from 20 to 30 inches of rainfall annually; over the Rockies themselves, semiarid conditions prevail; in the Great Basin, hemmed in by the Rockies on the east and the Sierra Nevadas on the west, more arid conditions predominate; to the west, over the Sierras and down to the seacoast, semiarid to the sub-humid conditions are again found.

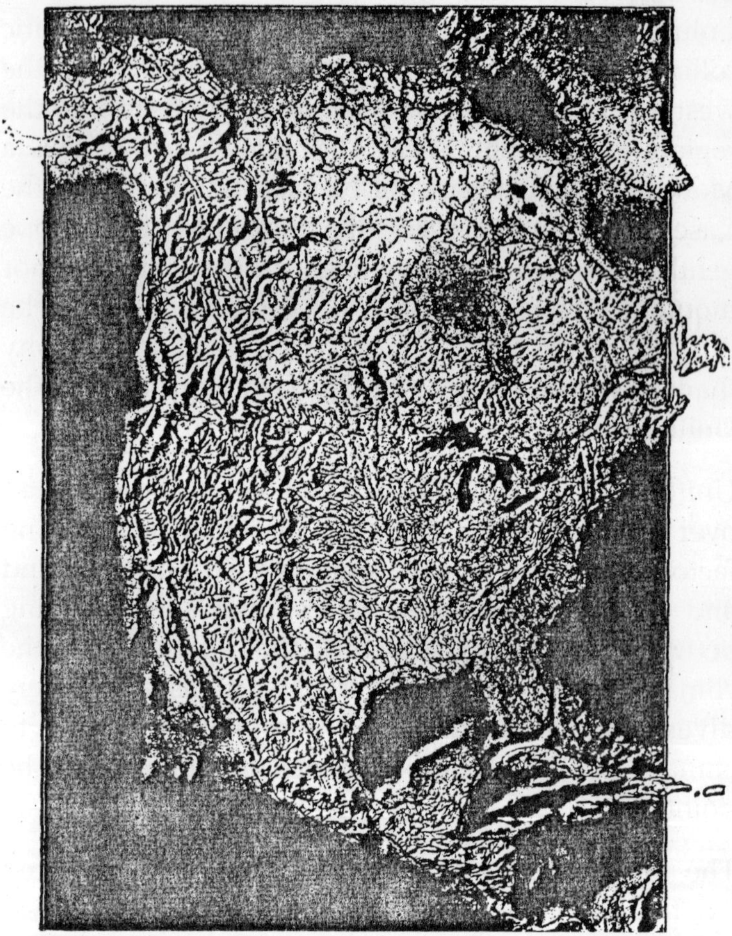

Fig. 7: Physical features of the dry-farm territory of the United Sates. Note the great variation of conditions. (After Tarr.)

Seasonal distribution of rainfall

It is doubtless true that the total annual precipitation is the chief factor in determining the success of dry-farming. However, the distribution of the rainfall throughout the year is also of great importance, and should be known by the farmer. A small rainfall, coming at the most desirable season, will have greater crop-producing power than a very much larger rainfall poorly distributed. Moreover, the methods of tillage to be employed where most of the precipitation comes in winter must be considerably different from those used where the bulk of the precipitation comes in the summer. The successful dry-farmer must known the average annual precipitation, and also the average seasonal distribution of the rainfall, over the land which the intends to dry-farm before he can safely choose his cultural methods.

With reference to the monthly distribution of the precipitation over the dry-farm territory of the United States, Henry of the United States Weather Bureau recognizes five distinct types; namely: (1) Pacific, (2) Sub-Pacific, (3) Arizona, (4) the Northern Rocky Mountain and Eastern Foothills, and (5) the Plains Type:-

(1) **The Pacific Type:** This type is found in all of the territory west of the Cascade and Sierra Nevada ranges, and also obtains in a fringe of country to the eastward of the mountain summits. The distinguishing characteristics of the Pacific types is a wet season, extending from October to March, and a practically rainless summer, except in

northern California and parts of Oregon and Washington. About half of the yearly precipitation comes in the months of the December, January, and February, the remaining half being distributed throughout the seven months – September, October, November, March, April, May and June".

(2) **Sub-Pacific Type:** The term 'Sub-Pacific' has been given to that type of rainfall which obtains over eastern Washington, Nevada, and Utah. The influences that control the precipitation of this region are much similar to those that prevail west of the Sierra Nevada and Cascade ranges. There is not however, as in the eastern type, a steady diminution in the precipitation with the approach of spring, but rather a culmination in eth precipitation".

(3) **Arizona Type:** The Arizona Type, so called because it is more fully developed in that territory than elsewhere, prevails over Arizona, New Mexico, and a small portion of eastern Utah and Nevada. This type differs from all other in the fact that about 35 per cent of the rain falls in July and August. May and June are generally the months of lest rainfall".

(4) **The Northern Rocky Mountain and Eastern, Foot-hills Type:** This type is closely allied to that of the plains to the eastward, and the bulk of the rain falls in the foothills of the region in April and May; in Montana, in May and June".

(5) **The Plains Type:** This type embraces the greater part of the Dakotas, Nebraska, Kansas, Oklahoma, the Panhandle of Texas, and all the great corn and wheat states of the interior valleys. This region is characterized by a scant winter precipitation over the northern stats and moderately heavy rains during the growing season. The bulk of the rains comes in May, June, and July".

This classification, with the accompanying chart (Fig. 8), emphasizes the great variation in distribution of rainfall over the dry-farm territory of the country. West of the Rocky Mountains the precipitation comes chiefly in winter and spring, leaving the summers rainless; while east of the Rockies, the winters are somewhat rainless and the precipitation comes chiefly in spring and summer. The Arizona type stands midway between these types. This variation in the distribution of the rainfall requires that different methods be employed in storing and conserving the rainfall for crop production. The adaptation of cultural methods to the seasonal distribution of rainfall will be discussed hereafter.

Snowfall

Closely related to the distribution of the rainfall and the average annual temperature is the snowfall. Wherever, a relatively large winter precipitation occurs, the dry-farmer is benefited if it comes in the from of now. The fall-planted seeds are better protected by the snow; the evaporation is lower and is appears that the soil is improved by the annual covering of snow.

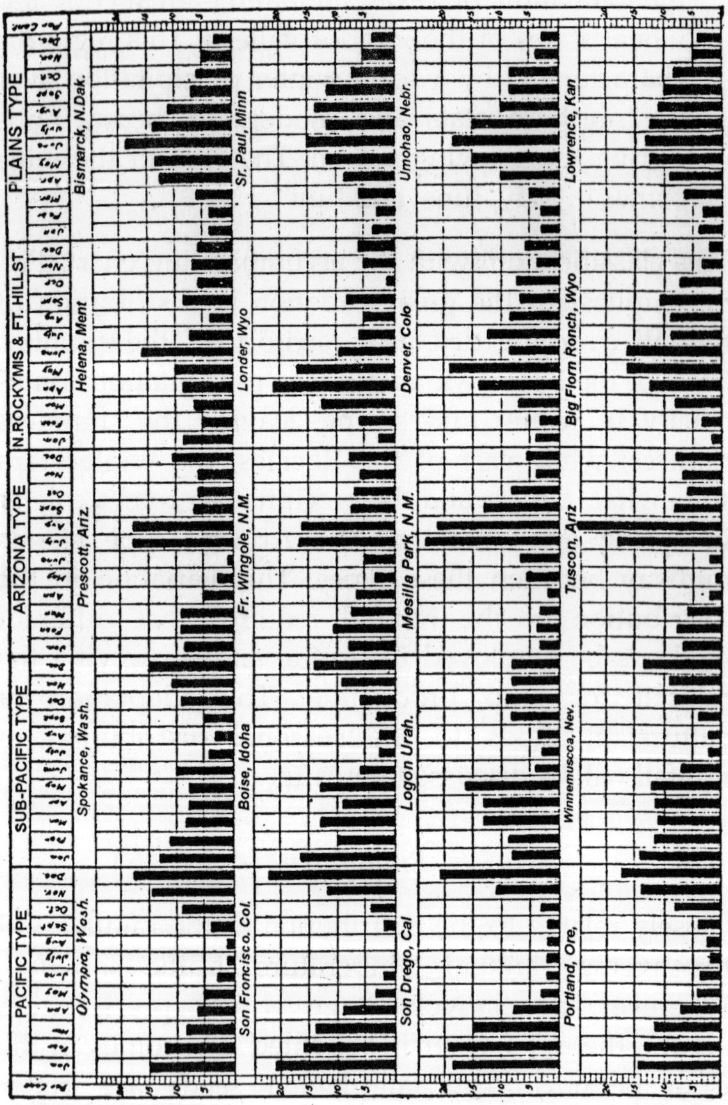

Fig. 8: The types of the distribution of the rainfall over the dry-farm territory of the United States. (After Henry).

In any case, the methods of culture are in a measure dependent upon the amount of snowfall and the length of time that it lies upon the ground.

Snow falls over most of the dry-farm territory, excepting the lowlands of California, the immediate Pacific coast, and other districts where the average annual temperature is high. The heaviest snowfall is in the intermountain district, from the west slope of the Sierra Nevadas to the east slope of the Rockies. The degree of snowfall on the agricultural lands is very variable and dependent upon local conditions. Snow falls upon all the high mountain ranges.

Temperature

With the exceptions of portions of California, Arizona, and Texas the average annual surface temperature of the dry-farm territory of the United States ranges from 40° to 55°F. The average is not far from 45°F. This places most of the dry-farm territory in the class of cold regions, though a small area on the extreme east border may be classed as temperate, and parts of California and Arizona as warm. The range in temperature from the highest in summer to the lowest in winter is considerable, but not widely different from other similar parts of the United States. The range is greatest in the interior mountainous districts, and lowest along the seacoast. The daily range of the highest and lowest temperatures for any one day is generally higher over dry-farm sections than over humid districts. In the Plateau regions of the semiarid country the average daily variation is from 30 to 35°F., while east of the Mississippi it is only about 20°F. This greater daily range is chiefly due to the clear skies and scant vegetation which facilitate excessive warming by day and cooling by night.

The important temperature question for the dry-farmer is whether the growing season is sufficiently warm and long to permit the maturing of crops. There are few places, even at high altitudes in the region considered, where the summer temperature is so low as to retard the growth of plants. Likewise, the first and last killing frosts are ordinarily so far apart as to allow an ample growing season. It must be remembered that frosts are governed very largely by local topographic features, and must be known from a local point of view. It is a general law that frosts are more likely to occur in valleys than no hillsides, owing to the downward drainage of the cooled air. Further, the danger of frost increases with the altitude. In general, the last killing frost in spring over the dry-farm territory varies from March 15 to May 29, and the first killing frost in autumn from September 15 to November 15. These limits permit of the maturing of all ordinary farm crops, especially the grain crops.

Relative humidity

At a definite temperature, the atmosphere can hold only a certain amount of water vapor. When the air can hold no more, it is said to be saturated. When it is not saturated, the amount of water vapor actually held by the air is expressed in percentages of the quantity required for saturation. A relative humidity of 100 per cent means that the air is saturated; of 50 per cent, that it is only one half saturated. The drier the air is, the more rapidly does the water evaporate into it. To the dry-farmer, therefore, the relative humidity or degree of dryness of the air is of very great importance. According to Professor Henry, the chief characteristics of the geographic distribution of relative humidity in the United States are as follows: -

Along the coasts there is a belt of high humidity at all seasons, the percentage of saturation ranging from 75 to 80 per cent.

Inland, from about the 70th meridian east-ward to the Atlantic coast, the amount varies between 70 and 75 per cent.

The dry region is in the Southwest, where the average annual value is not over 50 per cent. In this region are included Arizona, New Mexico, western Colorado, and the greater portion of both Utah and Nevada. The amount of annual relative humidity in the remaining portion of the elevated district, between the 100th meridian on the east to the Sierra Nevada and the Cascades on the west, varies between 55 and 65 per cent. In July, August, and September, the mean values in the South-west sink as low as 20 to 30 per cent, while along the Pacific coast districts they continue about 80 per cent the year round. In the Atlantic coast districts, and generally east from the Mississippi River, the variation from month to month is not great. April is probably the driest month of the year.

The air of the dry-farm territory, therefore, on the whole, contains considerably less than two thirds the amount of moisture carried by the air of the humid states. This mans that evaporation from plant leaves and soil surfaces will go on more rapidly in semiarid than in humid regions. Against this danger, which cannot be controlled, the dry-farmer must take special precautions.

Sunshine

The amount of sunshine in a dry-farm section is also of importance. Direct sunshine promotes plant growth, but at the same time it accelerates the evaporation of water from the soil. The whole dry-farm territory receives more sunshine than do the humid sections. In fact, the amount of sunshine may roughly be said to increase as the annual rainfall decreases. Over the larger part of the arid and semiarid sections the sun shines over 70 per cent of the time (Fig. 9).

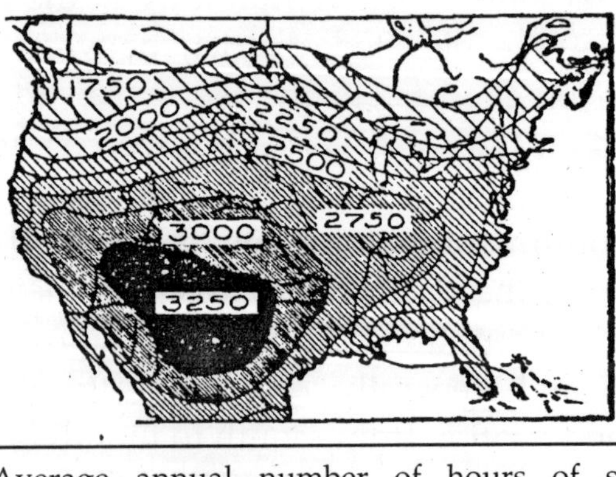

Fig. 9: Average annual number of hours of sunshine. (Cyclo. Am. Agr.)

Winds

The winds of any locality, owing to their moisture dissipating power, play an important part in the success of dry-farming. A persistent wind will offset much of the benefit of a heavy rainfall and careful cultivation. While great general laws have been formulated regarding the movements of the atmosphere, they are of minor value in judging the effect of wind on any farming district. Local observations, however, may enable the farmer to estimate the probable effect of the winds and thus to formulate proper cultural means of protection. In general, those living in a district are able to describe it without special observations as windy or quiet. In the dry-farm territory of the United States the one great region of relatively high and persistent winds is the Great Plains region east of the Rocky Mountains. Dry-farmers in that section will of necessity be obliged to adopt cultural methods that will prevent the excessive evaporation naturally induced by the unhindered wind, and the possible blowing of well-tilled fallow land.

Summary

The dry-farm territory is characterized by a low rainfall, averaging between 10 and 20 inches, the distribution of which falls into two distinct types: a heavy winter and spring with a light summer precipitation, and a heavy spring and summer with a light winter precipitation. Snow falls over most of the territory, but does not lie long outside of the mountain states. The whole dry-farm territory may be classed as temperate to cold; relatively high and persistent winds blow only over the Great Plains, though local conditions cause strong regular winds in many other places; the air is dry and the sunshine is very abundant. In brief, little water falls upon the dry-farm territory, and the climatic factors are of a nature to cause rapid evaporation.

In view of this knowledge, it is not surprising that thousands of farmers, employing, often carelessly, agricultural methods developed in humid sections, have found only hardships and poverty on the present dry-farm empire of the United States.

Drought

Drought is said to be the arch enemy of the dry-farmer, but few agree upon its meaning. For the purposes of this volume, drought maybe defined as a condition under which crops fail to mature because of an insufficient supply of water. Providence has generally been charged with causing droughts, but under the above definition. Man is usually the cause. Occasionally, relatively dry years occur, but they are seldom dry enough to cause crop failures if proper methods of farming have been practiced. There are four chief causes of drought:

(1) Improper or careless preparation of the soil;

(2) failure to store the natural precipitation in the soil;

(3) failure to apply proper cultural methods for keeping the moisture in the soil until needed by plants, and

(4) sowing too much seed for the available soil-moisture.

Crop failures due to untimely frosts, blizzards, cyclones, tornadoes, or hail may perhaps be charged to Providence, but the dry-farmer must accept the responsibility for any crop injury resulting from drought. A fairly accurate knowledge of the climatic conditions of the district, a good understanding of the principles of agriculture without irrigation under a low rainfall, and a vigorous application of these principles as adapted to the local climatic conditions will make dry-farm failures a rarity.

CHAPTER V

DRY-FARM SOILS

Important as is the rainfall in making dry-farming successful, it is not more so than the soils of the dry-farms. On a shallow soil, or on one penetrated with gravel streaks, crop failures are probable even under a large rainfall; but a deep soil of uniform texture, unbroken by gravel or hardpan, in which much water maybe stored, and which furnishes also an abundance of feeding space for the roots, will yield large crops even under a very small rainfall. Likewise, an infertile soil, though it be deep, and under a large precipitation, cannot be depended on for good crops; but a fertile soil, though not quite so deep, nor under so large a rainfall, will almost invariably bring large crops to maturity.

A correct understanding of the soil, from the surface to a depth of ten feet, is almost indispensable before a safe judgment can be pronounced upon the full dry-farm possibilities of a district. Especially is it necessary to known (a) the depth, (b) the uniformity of structure, and (c) the relative fertility of the soil, in order to plan an intelligent system of farming that will be rationally adapted to the rainfall and other climatic factors.

It is a matter of regret that so much of our information concerning the soils of the dry-farm territory of the United States and other counties has been obtained according to the methods and for the needs of humid countries, and that, therefore, the special knowledge of our arid and semiarid soils needed for the development of dry-farming is small and fragmentary. What is known to–day concerning the nature of arid soils and their relation to cultural processes under a scanty rainfall is due very largely to the extensive researches and voluminous writings of Dr. E. W. Hilgard, who for a generation was in charge of the agricultural work of the state of California. Future students of arid soils must of necessity rest their investigations upon the pioneer work done by Dr. Hilgard. The contents of this chapter are in a large part gathered from Hilgard's writings.

The formation of soils

"Soil is the more or less loose and friable material in which, by means of their roots, plants may or do find a foothold and nourishment, as well as other conditions of growth". Soil is formed by a complex process, broadly known as *weathering,* from the rocks which constitute the earth's crust. Soil is in fact only pulverized and altered rock. The forces that produce soil from rocks are of two distinct classes *physical* and *chemical*. The physical agencies of soil production merely cause a pulverization of the rock; the chemical agencies, on the other hand, so thoroughly change the essential nature of the soil particles that they are no longer like the rock from which they were formed.

Of the physical agencies, *temperature changes* are first in order of time, and perhaps of first importance. As the heat of the day increases, the rock expands, and as the cold night approaches, contracts. This alternate expansion and contraction, in time, cracks the surfaces of the rocks. Into the tiny crevices thus formed water enters from the falling snow or rain. When winter comes, the water in these cracks freezes to ice, and in so doing expands and widens each of the cracks. As these processes are repeated from day to day, from year to year, and from generation to generation, the surfaces of the rocks crumble. The smaller rocks so formed are acted upon by the same agencies, in the same manner, and thus the process of pulverization goes on.

It is clear, then, that the second great agency of soil formation, which always acts in conjunction with temperature changes, is *freezing water*. The rock particles formed in this manner are often washed down into the mountain valleys, there caught by great rivers, ground into finer dust, and at length deposited in the lower valleys. *Moving water* thus becomes another physical agency of soil production. Most of the soils covering the great dry-farm territory of the Untied States and other countries have been formed in this way.

In places, glaciers moving slowly down the canons crush and grind into powder the rock over which they pass and deposit it lower down as soils. In other places, where strong winds blow with frequent regularity, sharp soil grains are picked up by the air and hurled against the rocks, which, under this action, are carved into fantastic forms. In still other places, the strong winds carry soil over long distances to be mixed with other soils. Finally on the seashore the great waves dashing against the rocks of the coast line, and rolling the mass of pebbles back and forth, break and pulverize the rock until soil is formed. *Glaciers, winds* and *waves* are also, therefore, physical agencies of soil formation.

It may be noted that the result of the action of all these agencies is to from a rock powder, each particle of which preserves the composition that it had while it was a constituent part of the rock. It may further be noted that the chief of these soil-forming agencies act more vigorously in arid than in humid sections. Under the cloudless sky and dry atmosphere of regions of limited rainfall, the daily and seasonal temperature changes are much greater than in sections of greater rainfall. Consequently the pulverization of rocks goes on most rapidly in dry-farm districts. Constant heavy winds, which as soil formers are second only to temperature changes and freezing water, are also usually more common in arid than in humid countries. This is strikingly shown, for instance, on the Colorado desert and the Great Plains.

The rock powder formed y the processes above described is continually being acted upon by agencies, the effect of which is to change its chemical composition. Chief of these agencies is *water*, which exerts a solvent action on all known substances. Pure water exerts a strong solvent action, but when it has been rendered impure by a variety of substances, naturally occurring, it s solvent action is greatly increased.

The most effective water impurity, considering soil formation, is the gas, *carbon dioxid*. This gas is formed whenever plant or animal substances decay, and is therefore found, normally, in the atmosphere and in soils. Rains or flowing water gather the carbon dioxide from the atmosphere and the soil; few natural waters are free from it. The hardest rock particles are disintegrated by carbonated water, while limestones, or rocks containing lime, are readily dissolved.

The result of the action of carbonated water upon soil particles is to render soluble, and therefore more available to plants, many of the important plant-foods. In this way the action of water, holding in solution carbon dioxid and other substances, tends to make the soil more fertile.

The second great chemical agency of soil formation is the *oxygen* of the air. Oxidation is a process of more or less rapid burning, which tends to accelerate the disintegration of rocks.

Finally, the *plants* growing in sols are powerful agents of soil formation. First, the roots forcing their way into the soil exert a strong pressure which helps to pulverize the soil grains; secondly, the acids of the plant roots actually dissolve the soil, and third, in the mass of decaying plants, substances are formed, among them carbon dioxid, that have the power of making soils more soluble.

It may be noted that moisture, carbon dioxid, and vegetation, the three chief agents inducing chemical changes in soils, are most active in humid districts. While, therefore, the physical agencies of soil formation are most active in arid climates, the same cannot be said of the chemical agencies. However, whether in arid or humid climates, the processes of soil formation above outlined, are essentially those of the "fallow" or resting-period given to dry-farm lands. The fallow lasts for a few months or a year, while the process of soil formation is always going on and has gone on for ages; the result in quality though not in quantity, is the same – the rock particles are pulverized and the plant-foods are liberated. It must be remembered in this connection that climatic differences may and usually do influence materially the character of soils formed from one and the same kind of rock.

Characteristics of arid soils

The net result of the soil-forming processes above described is a rock powder containing a great variety of sizes of soil grains intermingled with clay. The larger soil grains are called sand; the smaller, silt, and those that are so small that they do not settle from quite water after 24 hours are known as clay. Compare Fig. 10.

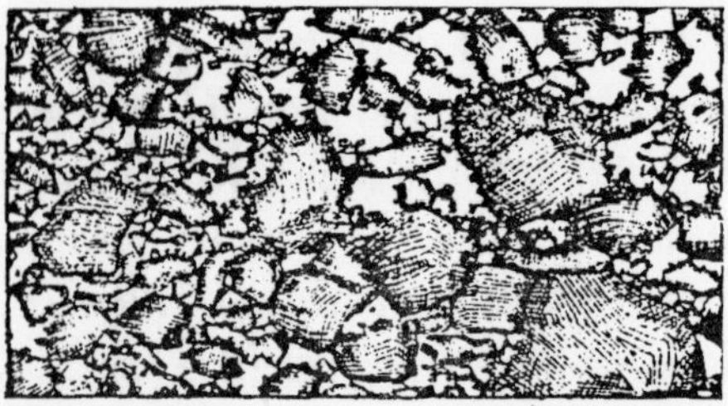

Fig. 10: Soil is a mixture of particles of very varying size.

Clay differs materially from sand and silt, not only in size of particles, but also in properties and formation. It is said that clay particles reach a degree of fineness equal to 1/2500 of an inch. Clay itself, when wet and kneaded, becomes plastic and adhesive and is thus easily distinguished from sand. Because of these properties, clay is of great value in holding together the larger soil grains in relatively large aggregates which give soils the desired degree of tilth. Moreover, clay is very retentive of water, gases, and soluble plant-foods, which are important factors in successful agriculture. Soils, in fact, are classified according to the amount of clay that they contain. Hilgard suggests the following classification:-

TABLE: 4

Very sandy soils	0.5 to 3 per cent clay
Ordinary sandy sols	3.0 to 10 per cent clay
Sandy loams	10.0 to 15 per cent clay
Clay loams	15.0 to 25 per cent clay
Clay soils	25.0 to 35 per cent clay
Heavy clay soils	35.0 per cent and over.

Clay may be formed from any rock containing some form of *combined silica* (quartz). Thus, granites and crystalline rocks generally, volcanic rocks, and shales will produce clay if subjected to the proper climatic conditions. In the formation of clay, the extremely fine soil particles are attacked by the soil water and subjected to deep-going chemical changes. In fact, clay represents the most finely pulverized and most highly decomposed and hence in a measure the most valuable portion of the soil. In the formation of clay, water is the most active agent, and under humid conditions its formation is most raid.

It follows that dry-farm soils formed under a more or less rainless climate contain less clay than do humid soils. This difference is characteristic, and accounts for the statement frequent made that heavy clay soils are not the best for dry-farm purposes. The fact is, that heavy clay soils are very rare in arid regions; if found at all, they have probably been formed under abnormal conditions, as in high mountain valleys, or under prehistoric humid climates.

Sand: The sand-forming rocks that are not capable of clay production usually consists of *uncombined silica* or quartz, which when pulverized by the soil-forming agencies give a comparatively barren soil. Thus it has come about that ordinarily a clayey soil is considered "strong" and a sandy soil "weak". Though this distinction is true in humid climates, where clay formation is rapid, it is not true in arid climates, where true clay is formed very slowly. Under conditions of deficient rainfall, soils are naturally less clayey, but as the sand and silt particles are produced from rocks which under humid conditions would yield clay, arid soils are not necessarily less fertile.

Experiments has shown that the fertility in the sandy soils of arid sections is as large and as available to plants as in the clayey soils of humid regions. Experience in the arid section of America, in Egypt, India, and other desert-like regions has further proved that the sands of the deserts produce excellent crops whenever water is applied to them. The prospective dry-farmer, therefore, need not be afraid of a somewhat sandy soil, provided it has been formed under arid conditions. In truth, a degree of sandiness is characteristic of dry-farm soils.

The *humus* content forms another characteristic difference between arid and humid soils. In humid regions plants cover the soil thickly; in arid regions they are branched scantily over the surface; in the former case the decayed remnants of generations of plants form a large percentage of humus in the upper soil; in the latter, the scarcity of plant life makes the humus content low. Further, under an abundant rainfall the organic matter in the soil rots slowly; whereas in dry warm climates the decay is very complete. The prevailing forces in all countries of deficient rainfall therefore tend to yield soils low in humus.

While the total amount of humus in arid soils is very much lower than in humid soils, repeated investigation has shown that it contains about 3½ times more nitrogen than is found in humus formed under an abundant rainfall. Owing to the prevailing sandiness of dry-farm soils, humus is not needed so much to give the proper tilth to the soil as in the humid countries where the content of clay is so much higher. Since, for dry-farm purposes, the nitrogen content is the most important quality of the humus, the difference between arid and humid soils, based upon the humus content, is not so great as would appear at first sight.

Soil and subsoil: In countries of abundant rainfall, a great distinction exists between the soil and the subsoil. The soil is represented by the upper few inches which are filled with the remnants of decayed vegetable matter and modified by plowing, harrowing, and other cultural operations. The sub-soil has been profoundly modified by the action of the heavy rainfall, which, in soaking through the soil, has carried with it the finest soil grains, especially the clay, into the lower soil layers.

In time, the subsoil has become more distinctly clayey then the topsoil. Lime and other soil ingredients have likewise been carried down by the rains and deposited at different depths in the soil or wholly washed away. Ultimately, this results in the removal from the topsoil of the necessary plant-foods and the accumulation in the subsoil as to make it difficult for roots and even air to penetrate it. The normal process of weathering or soil disintegration will then go on most actively in the topsoil, and the subsoil will remain unweathered and raw. This accounts for the well-known fact that in humid countries any subsoil that may have been plowed up is reduced to a normal state of fertility and crop production only after several years of exposure to the elements. The humid farmer, knowing this, is usually very careful not to let his plow enter the sub-soil to any great depth.

In the arid regions or wherever a deficient rainfall prevails, these conditions are entirely reversed. The light rainfall seldom completely fills the soil pores to any considerable depth, but it rather moves down slowly as a film, enveloping the soil grains. The soluble materials of the soil are, in part at least, dissolved and carried down to the lower limit of the rain penetration, but the clay and other fine soil particles are not moved downward to any great extent. These conditions leave the soil and subsoil of approximately equal porosity. Plant roots can then penetrate the soil deeply, and the air can move up and down through the soil mass freely and to considerable depths. As a result, arid soils are weathered and made suitable for plant nutrition to very great depths. In fact, in dry-farm regions there need be little talk about soil and subsoil, since the soil is uniform in texture and usually nearly so in composition, fro the top down to a distance of many feet.

Many soil sections 50 or more feet in depth are exposed in the dry-farming territory of the United States, and it has often been demonstrated that the subsoil to any depth is capable of producing, without further weathering, excellent yields of crops. This granular, permeable structure, characteristic of arid soils, is perhaps the most important single quality resulting from rock disintegration under arid conditions. As Hilgard remarks, it would seem that the farmer in the arid region owns from three to four farms, one above the other, as compared with the same acreage in the eastern states.

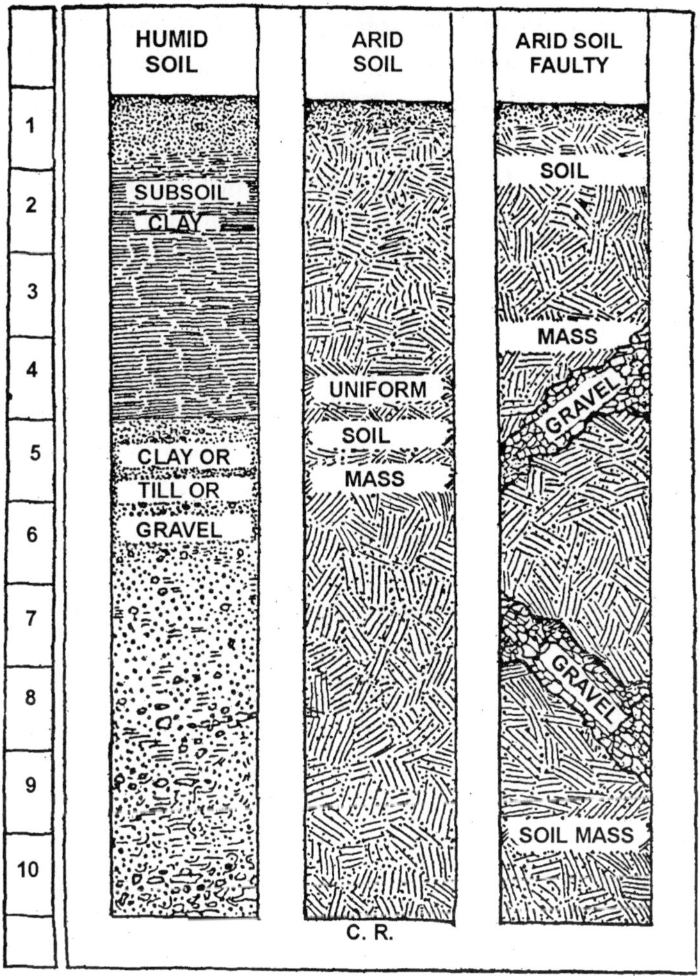

Fig. 11: Difference in structure between humid and arid soil. The **third** section, though arid, is questionable for dry-farming. The graved streaks break the continuity of the soil mass. (Adapted from Hilgard.).

This condition is of the greatest importance in developing the principles upon which successful dry-farming rests. Further, it may be said that while in the humid East the farmer must be extremely careful not to turn up with his plow too much of the inert subsoil, no such fear need possess the western farmer. On the contrary, he should use his utmost endeavor to plow as deeply as possible in order to prepare the very best reservoir for the falling waters and a place for the development of plant roots. Figure 11 shows graphically the difference existing between the soils of the arid and humid regions.

Gravel seams: It need be said, however, that in a number of localities in the dry-farm territory the soils have been deposited by the action of running water in such a way that the otherwise uniform structure of the soil is broken by occasional layers of loose gravel. While this is not a very serious obstacle to the downward penetration of roots, it is very serious in dry-farming, since any break in the continuity of the soil mass prevents the upward movement of water stored in the lower soil depths. The dry-farmer should investigate the soil which he intends to sue to a depth of at least 8 to 10 feet to make sure, first of all, that he has a continuous soil mass, not too clayey in the lower depths, nor broken by deposits of gravel.

Hardpan: Instead of the heavy cay subsoil of humid regions, the so-called hardpan occurs in regions of limited rainfall. The annual rainfall, which is approximately constant, penetrates from year to year very nearly to the same depth. Some of the lime found so abundantly in arid soils is dissolved and worked down yearly to the lower limit of the rainfall and left there to enter into combination with other soil ingredients. Continued through long periods of time this results in the formation of a layer of calcareous material at the average depth to which the rainfall has penetrated the soil. Not only is the lime thus carried down, but the finer particles are carried down in like manner. Especially where the soil is poor in lime is the clay worked down to form a somewhat clayey hardpan. A hardpan formed in such a manner is frequently a serious obstacle to the downward movement of the roots, and also prevents the annual precipitation from moving down far enough to be beyond the influence of the sunshine and winds. It is fortunate, however, that in the great majority of instances this hardpan gradually disappears under the influence of proper methods of dry-farm tillage. Deep plowing and proper tillage, which allow the rain waters to penetrate the soil, gradually break up and destroy the hardpan, even when it is 10 feet below the surface. Nevertheless, the farmer should make sure whether or not the hardpan does exist in the soil and plan his methods accordingly. If a hardpan is present, the land must be fallowed more carefully every other year, so that a large quantity of water may be stored in the soil to open and destroy the hardpan.

Of course, in arid as in humid countries, it often happens that a soil is underlaid, more or less near the surface, by layers of rock, marl deposits, and similar impervious or hurtful substances. Such deposits are not to be classed with the hardpans that occur normally wherever the rainfall in small.

Leaching: Fully as important as any of the differences above outlined are those which depend definitely upon the leaching power of a heavy rainfall. In countries where the rainfall is 30 inches or over, and in many places where the rainfall is considerably less, the water drains through the soil into the standing ground water. There is, therefore, in humid countries, a continuous drainage through the soils after every rain, and in general there is a steady downward movement of soil-water throughout the year. As is clearly shown by the appearance, taste, and chemical composition of drainage waters, this process leaches out considerable quantities of the soluble constituents of the soil.

When the soil contains decomposing organic matter, such as roots, leaves, stalks, the gas carbon dioxid is formed, which, when dissolved in water, forms a solution of great solvent power. Water passing through well-cultivated soils containing much humus leaches out very much more material than pure water could do. A study of the composition of the drainage waters from soils and the waters of the great rivers shows that immense quantities of soluble soil constituents are taken out of the soil in countries of abundant rainfall. These materials ultimately reach the ocean, where they are and have been concentrated throughout the ages. In short, the saltiness of the ocean is due to the substances that have been washed from the soils in countries of abundant rainfall.

In arid regions, on the other hand, the rainfall penetrates the soil only a few feet. In time, it is returned to the surface by the action of plants or sunshine and evaporated into the air. It is true that under proper methods of tillage even the light rainfall of arid and semiarid regions may be made to pass to considerable soil depths, yet there is little if any drainage of water through the soil into the standing ground water. The arid regions of the world, therefore, contribute proportionately a small amount of the substances which make up the salt of the sea.

Alkali soils: Under favorable conditions it sometimes happens that the soluble materials, which would normally be washed out of humid soils, accumulate to so large a degree in arid soils as to make the lands unfitted for agricultural purposes. Such lands are called alkali lands. Unwise irrigation in arid climates frequently produces alkali spots, but many occur naturally. Such soils should not be chosen for dry-farm purposes, for they are likely to give trouble.

Plant-food content: This condition necessarily leads at once to the suggestion that the soils from the two regions must differ greatly in their fertility or power to produce and sustain plant life. It cannot be believed that the water-washed soils of the east retain as much fertility as the dry soils of the West. Hilgard has made a long and elaborate study of this somewhat difficult question and has constructed a table showing the composition of typical soils of representative states in the arid and humid regions. The following table shows a few of the average results obtained by him:-

TABLE: 5

Source of soil	Number of samples analyzed	Partial percentage composition						
		Insoluble residue	Soluble silica	Alumina	Lime	Potash	Phosphoric acid	Humus
Humid region	696	84.17	4.04	3.66	0.13	0.21	0.12	1.22
Arid region	573	69.16	6.71	7.61	1.43	0.67	0.16	1.13

Soil chemists have generally attempted to arrive at a determination of the fertility of soil by treating a carefully selected and prepared sample with a certain amount of acid of definite strength. The portion which dissolves under the influences of acids has been looked upon as a rough measure of the possible fertility of the soil.

The column headed "Insoluble Residue" shown the average proportions of arid and humid soils which remain undissolved by acids. It is evident at once that the humid soils are much less soluble in acids than arid soils, the difference being 84 to 69. Since the only plant-food in sols that maybe used for plant production is that which is soluble, it follows that it is safe to assume that arid sols are generally more fertile than humid soils. This is borne out by a study of the constituents of the soil. For instance, potash, one of the essential plant foods ordinarily present in sufficient amount, is found in humid soils to the extent of 0.21 per cent, or over three times as much. Phosphoric acid, another of the very important plant-foods, is present in arid soils in only slightly higher quantities than in humid soils. This explains the somewhat well-known fact that the first fertilizer ordinarily required by arid soils is some form of phosphorus.

The difference in the chemical composition of arid and humid soils is perhaps shown nowhere better than in the lime content. There is nearly eleven times more lime in arid than in humid soils. Conditions of aridity favor strongly the formation of lime, and since there is very little leaching of the soil by rainfall, the lime accumulates in the soil.

The presence of large quantities of lime in arid soils has a number of distinct advantages, among which the following are most important:

(1) It prevents the sour condition frequently presenting humid climates, where much organic material is incorporated with the soil.

(2) When other conditions are favorable, it encourages bacterial life, which, as is now a well-known fact, is an important factor in developing and maintaining soil fertility.

(3) By somewhat subtle chemical changes it makes the relatively small percentages of other plant-foods, notably phosphoric acid and potash, more available for plant growth.

(4) It aids to convert rapidly organic matter into humus which represents the main portion of the nitrogen content of the soil.

Of course, an excess of lime in the soil may be hurtful, though less so in arid than in humid regions. Some authors state that from 8 to 20 per cent of calcium carbonate makes a soil unfitted for plant growth. There are, however, a great many agricultural soils covering large areas and yielding very abundant crops which contain very much larger quantities of calcium carbonate. For instance, in the Sanpete Valley of Utah, one of the most fertile sections of the Great Basin, agricultural soils often contain as high as 40 per cent of calcium carbonate, without injury to their crop-producing power.

In the table are two columns headed "Soluble Silica" and "Alumina", in both of which it is evident that a very much larger per cent is found in the arid than in the humid soils. These soil constituents indicate the condition of the soil with reference to the availability of its fertility for plant use. The higher the percentage of soluble silica and alumina, the more thoroughly decomposed, in all probability, is the soil as a whole and the more readily can plants secure their nutrient from the soil. It will be observed from the table, as previously stated, that more humus is found in humid than in arid soils, though the difference is not so large as might be expected. It should be recalled, however, that the nitrogen content of humus formed under rainless conditions is many times larger than that of humus formed in rainy countries, and that the smaller per cent of humus in dry-farming countries is thereby offset.

All in all, the composition of arid soils is very much more favorable to plant growth than that of humid soils. As will be shown in Chapter IX, the greater fertility of arid soils is one of the chief reasons for dry-farming success. Depth of the soil alone does not suffice. There must be a large amount of high fertility available for plants in order that the small amount of water can be fully utilized in plant growth.

Summary of characteristics: Arid soils differ from humid soils in that they contain: less clay; more sand, but of fertile nature because it is derived from rocks that in humid countries would produce clay less humus, but that of a kind which contains about 3½ times more nitrogen than the humus of humid soils; more lime, which helps in a variety of ways to improve the agricultural value of soils; more of all the essential plant-foods, because the leaching by downward drainage is very small in countries of limited rainfall.

Further, arid soils show no real difference between soil and subsoil; they are deeper and more permeable; they are more uniform in structure; they have hardpans instead of clay subsoil, which, however, disappear under the influence of cultivation; their subsoils to a depth of ten feet or more are as fertile as the topsoil, and the availability of the fertility is greater. The failure to recognize these characteristic differences between arid and humid soils has been the chief cause for many crop failures in the more or less rainless regions of the world.

This brief reviews shows that, everything considered, arid soils are superior to humid soils. In ease of handling, productivity, certainty of crop-lasting quality, they far surpass the soils of the countries in which scientific agriculture was founded. As Hilgard has suggested, the historical datum that the majority of the most populous and powerful historical peoples of the world have been located on soils hat thirst for water, may find its explanation in the intrinsic value of arid soils. From Babylon to the United States is a far cry; but it is one that shouts to the world the superlative merits of the soil that begs for water. To learn how to use the "desert" is to make it "blossom like the rose".

Soil divisions

The dry-farm territory of the Untied States maybe divided roughly into five great soil districts, each of which includes a great variety of soil types, most of which are poorly known and mapped. These districts are:-

1. Great Plains district.
2. Columbia River district.
3. Great Basin district.
4. Colorado River district.
5. California district.

Great Plains district: On the eastern slope of the Rocky Mountains, extending eastward to the extreme boundary of the dry-farm territory, are the soils of the High Plains and the Great Plains. This vast soil district belongs to the drainage basin of the Missouri, and includes North and South Dakota, Nebraska, Kansas, Oklahoma, and parts of Montana, Wyoming, Colorado, New Mexico, Texas, and Minnesota. The soils of this district are usually of high fertility. They have good lasting power, though the effect of the higher rainfall is evident in their composition. Many of the distinct types of the plains soils have been determined with considerable care by Snyder and

Lyon, and may be found described in Bailey's "Cyclopedia of America Agriculture", Vol. I.

Columbia River district: The second great soil district of the dry-farming territory is located in the drainage basin of the Columbia River, and includes Idaho and the eastern two thirds of Washington and Oregon. The high plains of this soils district are often spoken of as the Palouse country. The soils of the western part of this district are of basaltic origin; over the southern part of Idaho the soils have been made from a somewhat recent lava flow which in many places is only a few feet below the surface. The soils of this district are generally of volcanic origin and very much alike. They are characterized by the properties which normally belong to volcanic soils; somewhat poor in lime, but rich in potash and phosphoric acid. They last well under ordinary methods of tillage.

The Great Basin: The third great soil district is included in the Great Basin, which covers nearly all of Nevada, half of Utah, and takes small portions out of Idaho, Oregon, and southern California. This basin has no outlet to the sea. Its rivers empty into great saline inland lakes, the chief of which is the Great Salt Lake. The sizes of these interior lakes are determined by the amounts of water flowing into them and the rates of evaporation of the water into the dry air of the region.

In recent geological times, the Great Basin was filled with water, forming a vast fresh-water lake known as Lake Bonneville, which drained into the Columbia River. During the existence of this lake soil materials were washed from the mountains into the lake and deposited on the lake bottom. When, at length, the lake disappeared, the lake bottom was exposed and is now the farming lands of the Great Basin district. The soils of this district are characterized by great depth and uniformity, and abundance of lime, and all the essential plant-foods with the exception of phosphoric acid, which, while present in normal quantities, is not unusually abundant. The Great Basin soils are among the most fertile on the American Continent.

Colorado River district: The fourth soil district lies in the drainage basin of the Colorado River. It includes much of the southern part of Utah, the eastern part of Colorado, part of New Mexico, nearly all of Arizona, and part of southern California. This district, in its northern part, is often spoken of as the High Plateaus. The soils are formed from the easily disintegrated rocks of comparatively recent geological origin, which themselves are said to have been formed from deposits in a shallow interior sea which covered a large part of the West. The rivers running through this district have cut immense canons with perpendicular walls which make much of this country difficult to traverse. Some of the soils are of an extremely fine nature, settling firmly and requiring considerable tillage before they are brought to a proper condition of tilth. In many places the soils are heavily charged with calcium sulphate, or crystals of the ordinary land plaster. The fertility of the soils, however, is high, and when they are properly cultivated, they yield large and excellent crops.

California district: The fifth soil district lies in California in the basin of the Sacramento and San Joaquin rivers. The soils are of the typical arid kind of high fertility and great lasting powers. They represent some of the most valuable dry-farm districts of the West. These soils have been studied in detail by Hilgard.

Dry-farming in the five districts: It is interesting to note that in all of these five great soil districts dry-farming has been tried with great success. Even in the Great Basin and the Colorado River districts, where extreme desert conditions often prevail and the rainfall is slight, it has been found possible to produce profitable crops without irrigation. It is unfortunate that the study of the dry-farming territory of the Untied Sates has not progressed far enough to permit a comprehensive and correct mapping of its soils. Our knowledge of this subject is, at the best fragmentary. We know, however, with certainty that the properties which characterize arid soils, as described in this chapter, are possessed by the soils of the dry-farming territory, including the five great districts just enumerated. The characteristics of arid soils increase as the rainfall decreases and other conditions of aridity increase. They are less marked as we go eastward or westward toward the regions of more abundant rainfall; that is to say, the most highly developed arid soils are found in the Great Basin and Colorado River districts the least developed are on the eastern edge of the Great Plains.

The judging of soils

A chemical analysis of a soil, unless accompanied by a large amount of other information, is of little value to the farmer. The main points in judging a prospective dry-farm are: the depth of the soil, the uniformity of the soil to a depth of at least 10 feet, the native vegetation, the climatic conditions as relating to early and late frosts, the total annual rainfall and its distribution and the kinds and yields of crops that have been grown in the neighborhood.

The depth of the soil is best determined by the use of an auger (Fig. 12). A simple soil auger is made from the ordinary carpenter's auger, 1½ to 2 inches in diameter, by lengthening its shaft to 3 feet or more. Where it is not desirable to carry sectional augers, it is often advisable to have three augers made: one 3 feet, the other 6, and the third 9 or 10 feet in length. The short auger is used first and the others afterwards as the depth of the boring increases. The boring should be made in a large number of average places – preferably one boring or more on each acre if time and circumstances permit and the results entered on a map of the farm. The uniformity of the soil is observed as the boring progresses. If gravel layers exist, they will necessarily stop the progress of the boring. Hardpans of any kind will also be revealed by such an examination.

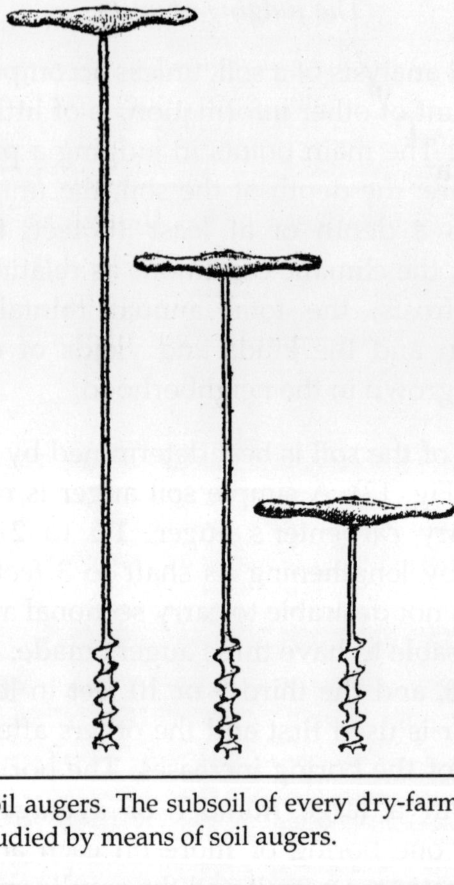

Fig. 12: Soil augers. The subsoil of every dry-farm should be studied by means of soil augers.

The climatic information must be gathered from the local weather bureau and from older residents of the section.

The native vegetation is always an excellent index of dry-farm possibilities. If a good stand of native grasses exists, there can scarcely be any doubt about the ultimate success of dry-farming under proper cultural methods. A healthy crop of sagebrush is an almost absolutely certain indication that farming without irrigation is feasible. The rabbit brush of the drier regions is also usually a good indication, though it frequently indicates a soil not easily handled. Greasewood, shadscale, and other related plants ordinarily indicate heavy clay soils, frequently charged with alkali. Such soils should be the last choice for dry-farming purposes, though they usually given good satisfaction under systems of irrigation. If the native cedar or other native trees grow in profusion, it is another indication of good dry-farm possibilities.

CHAPTER VI

THE ROOT SYSTEMS OF PLANTS

The great depth and high fertility of the soils of arid and semiarid regions have made possible the profitable production of agricultural plants under a rainfall very much lower than that of humid regions. To make the principles of this system fully understood, it is necessary to review briefly our knowledge of the root system of plants growing under arid conditions.

Function of roots

The roots serve at least three distinct uses or purposes: First, they give the plant a foothold in the earth; Secondly, they enable the plants to secure from the soil the large amount of water needed in plant growth, and, thirdly, they enable the plant to secure the indispensable mineral foods which can be obtained only form the soil. So important is the proper supply of water and food in the growth of a plant that, in a given soil, the crop yield is usually in direct proportion to the development of the root system. Whenever the roots are hindered in their development, the growth of the plant above ground is likewise retarded, and crop failure may result. The importance of roots is not fully appreciated because they are hidden form direct view. Successful dry-farming consists, largely, in the adoption of practices that facilitate a full and free development of plant roots. Were it not that the nature of arid soils, as explained in preceding chapters, is such that full root development is comparatively easy it would probably be useless to attempt to establish a system of dry-farming.

Kinds of roots

The root is the part of the plant that is found underground. It has numerous branches, twigs, and filaments. The root which first forms when the seed bursts is known as the primary root. From this primary root other roots develop, which are known as secondary roots. When the primary root grows more rapidly than the secondary roots, the so-called taproot, characteristic of lucern, clover, and similar plants, is formed. When, on the taproot grows slowly or ceases its growth, and the numerous secondary roots grow long, a fibrous root system results, which is characteristic of the cereals, grasses, con and other similar plants. With any type of root, the tendency of growth is downward; though under conditions that are not favorable of the downward penetration of the roots the lateral extensions may be very large and near the surface. (Figs. 22, 23).

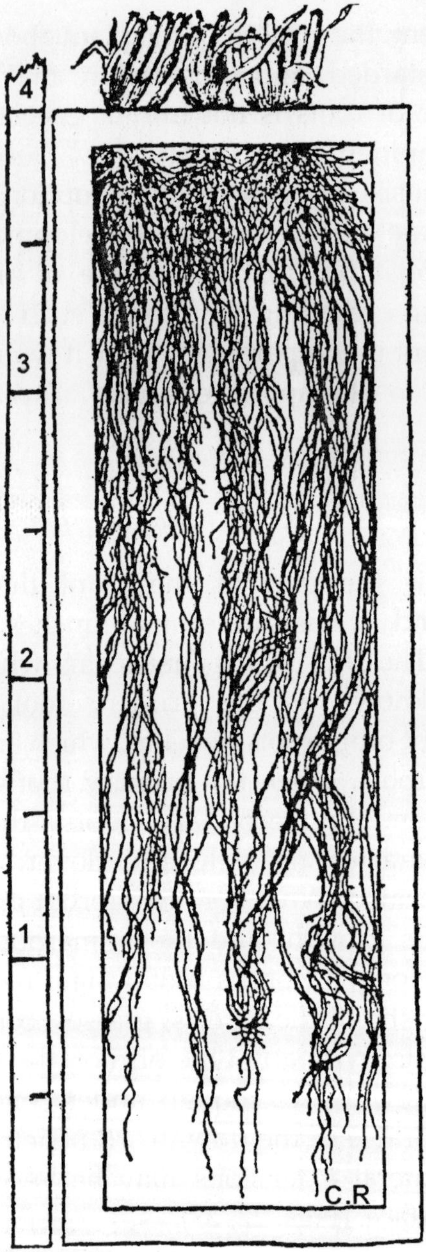

Fig. 13: Wheat roots.

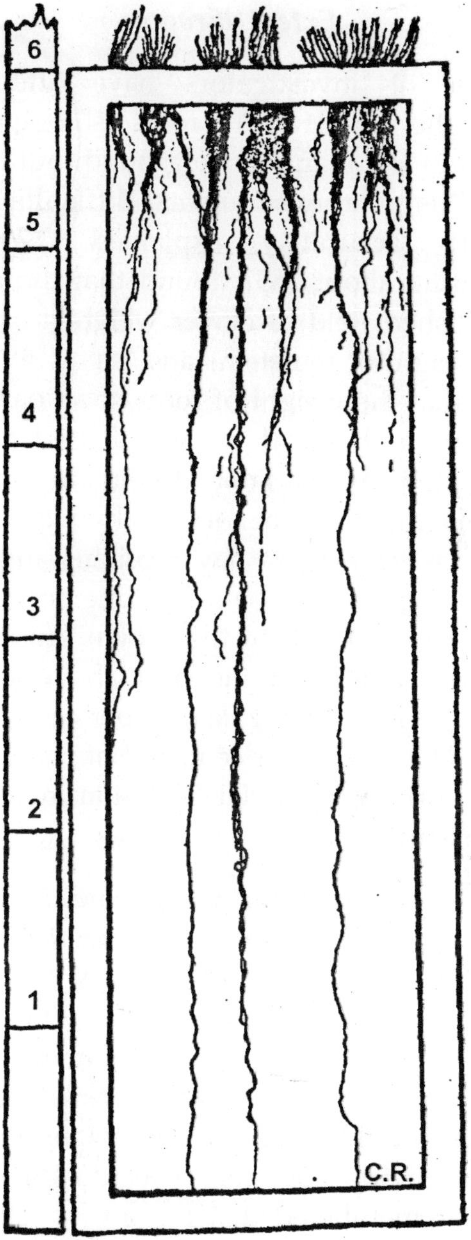

Fig. 14: Alfalfa roots.

Extent of roots

A number of investigators have attempted to determine the weight of the roots as compared with the weight of the plant above ground, but ht subject, because of its great experimental difficulties, has not been very accurately explained. Schumacher, experimenting about 1857, found that the roots of a well-established field of clover weighed as much as the total weight of the stems and leaves of the year's crop, and that the weight of roots of an oat crop was 43 per cent of the total weight of seed and straw. Nobbe, a few years later, found in one of his experiments that the roots of timothy weighed 31 per cent of the weight of the hay. Hosaeus, investigating the same subject about the same time, found that the weight of roots of one of the brown grasses was as great as the weight of the part above ground; of Serradella, 77 per cent; of flax, 34 per cent; of oats, 14 per cent; of barley, 13 per cent, and of peas, 9 percent. Sanborn, working at the Utah Station in 1893, found results very much same.

Although these results are not concordant, they show that the weight of the roots is considerable, in many cases far beyond the belied of those who have given the subject little or no attention. It may noted that on the basis of the figures above obtained, it is very probable that the roots in one acre of an average wheat crop would weigh in the neighborhood of a thousand pounds- possibly considerably more. It should be remembered that the investigation which yielded the preceding results were all conducted in humid climates and at a ti8me when the methods for the study of the root systems were poorly developed. The data obtained, therefore, represent, in all probability, minimum results which would be materially increased should the work be repeated now.

The relative weights of the roots and the stems and the leaves do not alone show the large quantity of roots; the total lengths of the roots are even more striking. The German investigator, Nobbe, in a laborious experiment conducted about 1867, added the lengths of all the fine roots form each of various plants. He founded that the total length of the roots, that is, the sum of the lengths of all the roots, of one wheat plant was about 298 feet, and that the total length of the roots of one plant of rye was about 385 feet. King, of Wisconsin, estimates that in one of his experiments, one corn plant produced in the upper 3 feet of soil 1452 feet of roots. These surprisingly large numbers indicate with emphasis the thoroughness with which the roots invade the soils. Figures 13-17 further give an idea of the degree to which roots fill the soil.

Depth of root penetration

The earlier root studies did not pretend to determine the depth to which roots actually penetrate the earth. In recent years, however, a number of carefully conducted experiments were made by the New York, Wisconsin, Minnesota, Kansas, Colorado, and especially the North Dakota stations to obtain accurate information concerning the depth to which agricultural plants penetrate soils. It is some what regrettable, for the purpose of dry-farming, that theses stages, with the exception of Colorado, are all in the humid or sub-humid area of the United states. Nevertheless, the conclusions drawn from the work are such that they may be safely applied in the development of principles of dry farming.

There is a general belief among farmers that the roots of all cultivated crops are very near the surface and that few reach a greater depth that one or two feet. The first striking result of the America investigations was that every crop, without exception, penetrates the soil deeper that was thought possible in earlier days. For example, it was found that corn roots penetrated fully four feet into the ground and that they fully occupied all of the soil to that depth.

On deeper and somewhat drier soils, corn roots went down as far as eight feet. The roots of the small grains, - wheat, oats, barley, penetrated the soil from four to eight or ten feet. Various perennial grasses rooted to a depth of four feet the first year; the next year, five and one half feet; no determination were made of the depth of the roots in later years., though it had undoubtedly increased. Alfalfa was the deepest rooted of all the crops studied by the American stations. Potato roots filled the soil fully to a depth of three feet; sugar beets to a depth of nearly four feet.

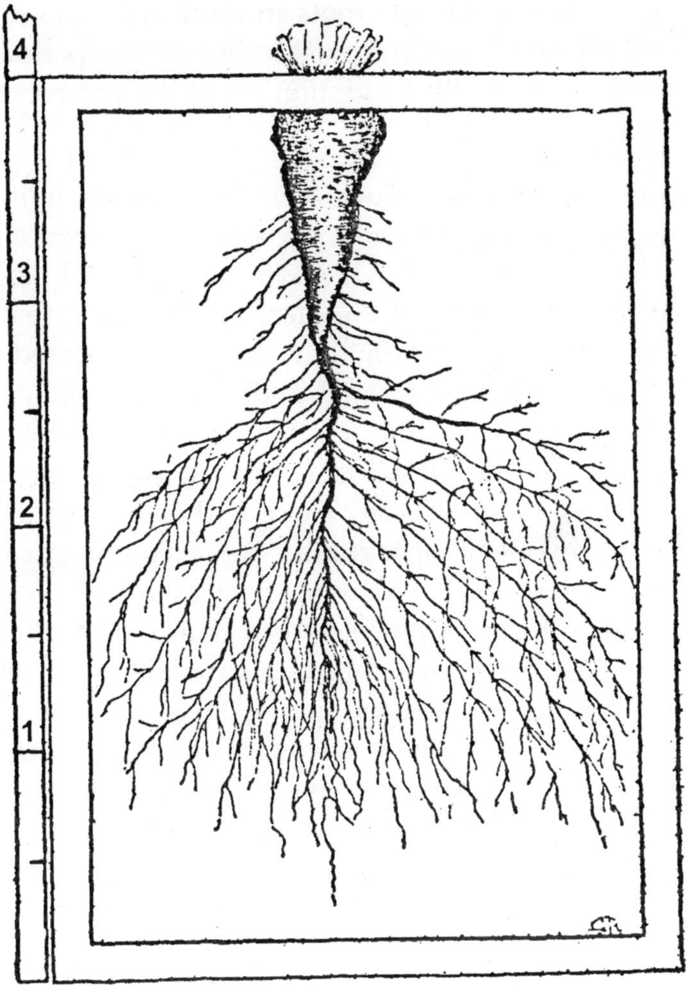

Fig. 15: Sugar-beet roots.

In every case, under conditions prevailing in the experiments, which did not have in mind the forcing of the roots down to extraordinary depths, it seemed that the normal depth of the root of ordinary field corps was from three to eight feet. Sub soiling and deep lowing enable the roots to go deeper into soil. This work has been confirmed in ordinary experience until there can be little question about the accuracy of the results.

Almost all of these results were obtained in humid climates on humid soils, somewhat shallow, and under laid by a more or less infertile subsoil. In fact, they were obtained under conditions really unfavorable to plant growth. It has been explained in Chapter V that soils formed under arid or semi arid conditions are uniformly deep and porous and that the fertility of the subsoil is, in most cases, practically as great as of the topsoil. There is therefore, in arid soils, an excellent opportunity for a comparatively easy penetration of the roots to great depths and, because of the available fertility, a chance throughout the whole of the subsoil for sample root development. Moreover, the porous condition of the soil permits the entrance of air, which helps to purify the soil atmosphere and thereby to make the conditions more favorable for root development. Consequently it is to be expected that in arid region, roots will ordinarily to a much grater depth that in humid regions.

It is further to be remembered that roots are in constant search of food and water and are likely to develop in the directions where there is the greatest abundance of these materials. Under systems of dry-farming the soil water is stored more or less uniformly to considerable depths- ten feet or more- and in most cases the percentage of moisture in. the spring and summer is as large or larger some feet below the surface than in the upper two feet.

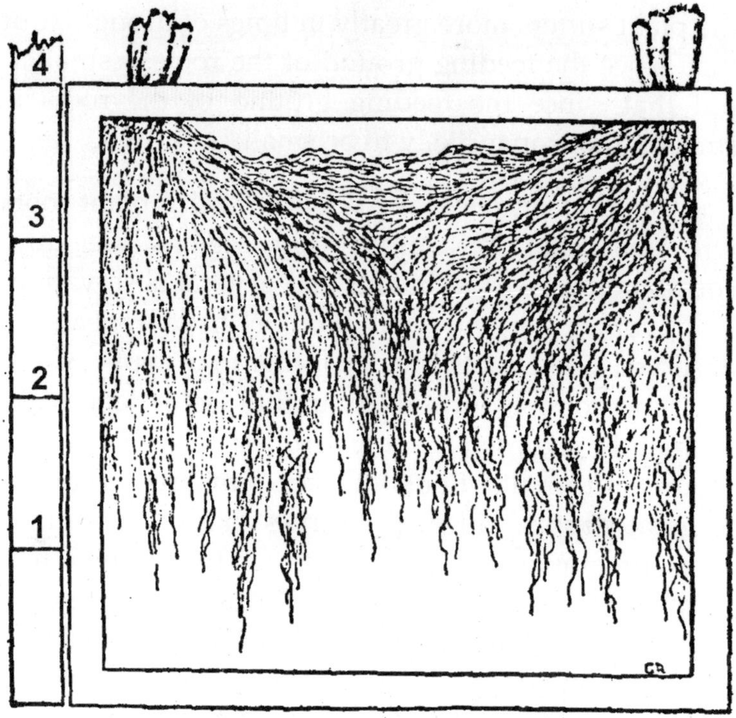

Fig. 16: Corn roots.

The tendency of the root is then, to move downward to depths where there is a larger supply of water. Especially is this tendency increased by the available soil fertility found throughout the whole depth of the soil mass.

Its has been argued that in many of the irrigated sections the roots do not penetrate the soil to great depths. This is true, because by the parent wasteful methods of irrigation the plant receives so much water at such untimely seasons that the roots acquire the habit of feeding very near the surface where the water is so lavishly applied. This means not only that the plant suffers more greatly in times of drought, but that, since the feeding ground of the roots is smaller, but that, since the feeding ground of the roots is smaller, the crop is likely to be small.

These deductions as to the depth to which plant roots will penetrate the soil in arid regions are fully corroborated by experiments and general observation. The workers of the Utah Station have repeatedly observed plant roots on dry-farms to a depth of ten feet. Lucern roots form thirty to fifty feet in length are frequently exposed in the gullies formed by the mountain torrents. Roots of trees, similarly, go down to great depths. Hilgard observes that he has found roots of grapevines at a depth of twenty-two feet below the surface, and quotes Aughey as having found roots of the native Shepherdia in Nebraska to a depth of fifty feet. Hilgard further declares that in California fibrous-rooted plants, such as wheat and barley, may descend in sandy soils from four to seven feet. Orchard trees in the arid West, grown properly, are similarly observed to send their roots down to great depths. In fact, it has become a custom in many arid regions where the soils are easily penetrable to say that the root system of a tree corresponds in extent and branching to the part of the tree above ground.

Now, it is to be observed that, generally, plants grown in dry climates send their roots straight down into the soil; whereas in humid climates, where the topsoil is quite moist and the subsoil is hard, roots branch out laterally and fill the upper foot or two of the soil this difference is made clear by the illustrations here with produced (Fig. 17). A great deal has been said and written about the danger of deep cultivation, because it tends to injure the roots that feed near the surface. However true this may be in humid countries as is often declared. True, deep cultivation, especially when performed near the plant or tree, destroys the surface feeding rots, but this only tends to compel the deeper lying roots to make better use of the subsoil.

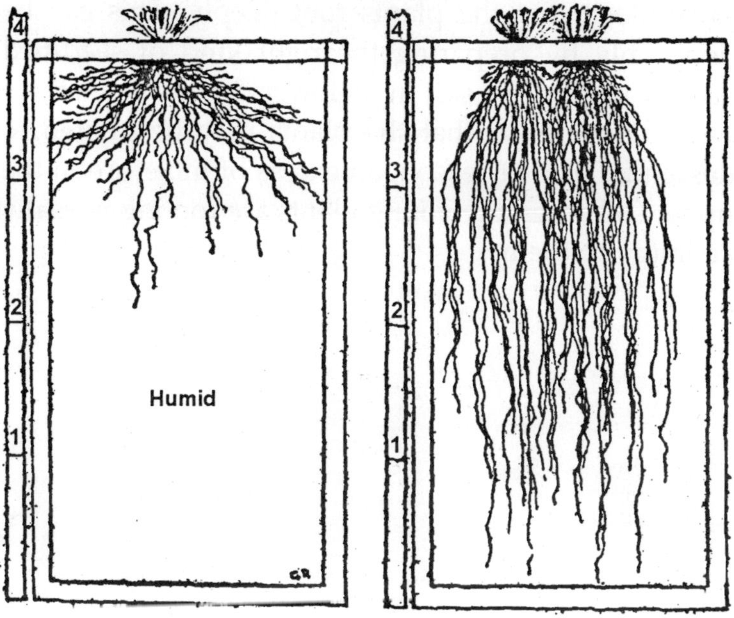

Fig. 17: Difference in root systems under humid and arid conditions.

When, as in arid regions, the subsoil is fertile and furnishes a sufficient amount of water, destroying the surface roots is no handicap whatever. On the contrary, in times of drought, the deep-lying roots feed and drink at their leisure far form the host sun or withering winds, while the plants with shallow roots wither and die or are so seriously injured as to produce an inferior crop. Therefore, in the system of dry-farming as developed in this volume it must be understood that so far as the farmer has power, the roots must be driven downward into the soil, and that no injury needs to be apprehended from deep and vigorous cultivation.

One of the chief attempts of the dry-farmers must be to see to it that the plants root deeply. This can be done only by preparing the right kind of seed-bed and by having the soil in its lower depths well stores with moisture, so that the plants may be invited to descend. For that reason, an excess of moisture in the upper soil when the young plants are rooting in really an injury to them.

CHAPTER VII

STORING WATER IN THE SOIL

The large amount of water required for the production of plant substance is taken from the soi8l by the roots. Leaves and stems do not absorb appreciable quantities of water. The scanty rainfall of dry-farm districts or the more abundant precipitation of humid regions must, therefore, be made to enter the soil in such a manner as to be readily available as soil moisture to the roots at the right periods of plants growth.

In humid countries, the rain falls during the growing season is looked upon, and very properly, as the really effective factor in the production of large crops. The roots systems of plants grown under such humid conditions are near the surface, ready to absorb immediately that rains that fall, even if they do not soak deeply into the soil. As has been shown in Chapter IV, it is only over a small portion of the dry farm territory that the bulk of the precipitation occurs during the growing season. Over a large portion of the arid and the semi arid region the summers are almost rainless and the bulk of the precipitation comes in the winter, late fall, or early spring when plants are not growing. If the rains that fall during the growing season are indispensable in crop production, the possible area to be reclaimed by dry farming will be greatly limited. Even when much of the total precipitation comes in summer, the amount in dry-farming depends chiefly upon the success with which the rains that fall during any season of the year may be stored and kept in the soil until needed by plants in their growth. The fundamental operations of dry-farming include a soil treatment which enables the largest proportion of the annul precipitation to be stored in the soil. For this purpose, the deep, somewhat porous soils, characteristic of arid regions are unusually well adapted.

Always demonstration

An important and unique demonstration of the possibility of bringing crops to maturity on the moisture stored in the soil at the time of planting has been made by always (Fig. 18). Cylinders of galvanized iron, 6 feet long, were filled with soil as nearly as possible in its natural position and condition. Water was added until seepage began, after which the excess was allowed to drain away. When the seepage had closed, the cylinders were entirely closed except at the surface. Sprouted grains of spring wheat were placed in the moist surface soil, and 1 inch of dry soil added to the surface to prevent evaporation. No more water was added; the air of the greenhouse was kept as dry as possible.

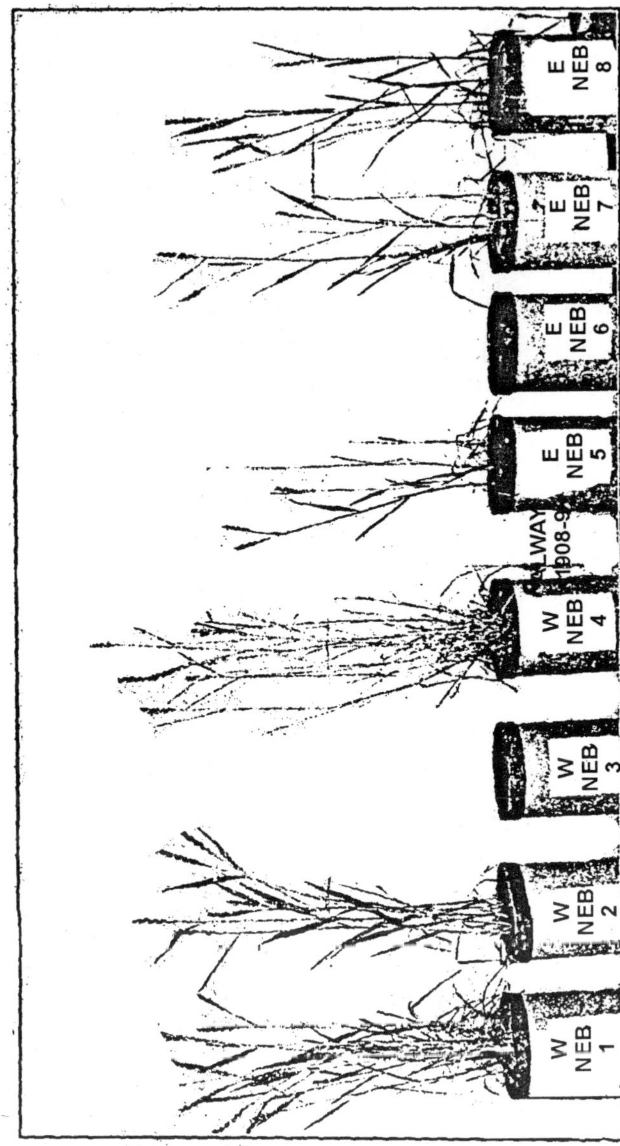

Fig. 18: Always experiment. The grain ripened with the water in the soil at planting time. No water was added after germination. Nebraska Experiment Station.

The wheat developed normally. The first ear was ripe in 132 days after planting and the last in 143 days the three cylinders of soil from semiarid western Nebraska produced 37.8 grams of straw and 29 ears, containing 415 kernels of soil form humid eastern Nebraska produced only 11.2 grams of straw and 13 ears containing 114 kernels, weighing 3 grams. This experiment shows conclusively that rains are not needed during the growing season, if the soil is well filled with moisture at seedtime, to bring crop to maturity.

What becomes of the rainfall?

The water that falls on the land is disposed of in three ways: First, under ordinary conditions, a large portion runs off without entering the soil; secondly, portion enters the soil, but remains near the surface, and is rapidly evaporated back into the air; and, thirdly, a portion enters the lower soil layers from which it is removed at later periods by several distinct processes. The run-off is usually large and is a serious loss, especially in dry farming regions, where; the absence of luxuriant vegetation, the somewhat hard, sun-baked soils and the numerous drainage channels, formed by successive torrents, combine to furnish the rains with an easy escape into the torrential rivers. Persons familiar with arid conditions know how quickly the narrow box canons, which often drain thousands of square miles, are filled with roaring water after a comparatively light rainfall.

The run-off

The proper cultivation of the soil diminishes very greatly the loss due to run-off, but even on such soils the proportion may often be very great. Farrel observed at one of the Utah stations that during a torrential rain – 2.6 inches in 4 hours- the surface of the summer fallowed plats was packed so solid that only one fourth inch, or less that n one tenth of the whole amount soaked into the soil while on neighboring stubble field, which offered greater hindrance to the run-off, 1½ inches or about 60 per cent were absorbed.

It is not possible under any condition to prevent the run-off altogether, although it can usually be reduced exceedingly. It is a common dry-farm custom to plow along the slopes of the farm instead of plowing up and down them. When this is done, the water which runs down the slopes is caught by the succession of furrows and in that way the run off is diminished. During the fallow season the disk and smoothing harrows are run along the hillsides for the same purpose and with results that are nearly always advantageous to the dry farmer. Of necessity, each man must study his own farm in order to devise methods that will prevent the run – off.

The structure of soils

Before examining more closely the possibility of storing water in soils a brief review of the structure of soils is desirable. As previously explained, soil is essentially a mixture of disintegrated rock and the decomposing remains of plants. The rock particles which constitute the major portion of soils vary greatly in size of the smallest. The following table shows the limits of sizes and the names used to designate them:-

TABLE: 6
NAMES AND SIZES OF THE SOIL PARTICLES

Name	Diameters In Millimeters	Numbers In One Lineal Inch	Number In One Cubic Inch
Sand	0.5-0.03	50-833	125,000-578,009, 537
Silt	0.03-0.001	833-25,000	578,009,537-15,625,000,000,000
Clay	Below 0.001	More than	More than
		25,000	15,625,000,000,000,

It will be observed that it would take 50 of the coarsest sand particles and 25,000 of the finest silt particles, to form one lineal inch. The clay particles are often smaller and of such a nature that they can not be accurately measured. The total number of soil particles in even a small quantity of cultivated soil is far beyond the ordinary limits of thought, ranging from 125,000 particles of coarse sand to 15,625,000,000,000 particles of the finest silt in one cubic inch. In other s words, if all the particles in one cubic inch of soil consisting of fine silt were placed side by side, they would form a continuous chain over a thousand miles long. The farmers, when he tills the soil deals with countless numbers of individual soil grains, far surpassing the understanding of the human mind. It is the immense number of constituent soil particles that gives to the sol many of its most valuable properties.

It must be remembered that no natural soil is made up of particles all of which are of the same size; all sizes, form the coarsest sand to the finest clay, are usually present (Fig. 10). These particles of all sizes are not arranged in the soil in a regular, orderly way; they are not place side by side with geometrical regularity; they are rather jumbled together in every possible way. The larger sand grains touch and form comparatively large interstitial spaces into which the finer silt and clay grains filter. Then, again the clay particles, which have cementing properties, blind, as it were, one particle to another. A sand grain may have attached to it hundreds, or it may be thousands of the smaller silt grains; or a regiment of smaller soils grain by cementing power of the clay. Further, in the presence of lime and similar substances these complex soils grains are grouped into yet larger and more complex groups. The beneficial effect of lime is usually due to this power of grouping untold numbers of soil culture the individual soil grains are thus grouped into large clusters, the soil is said to be in good tilth. Anything that tends to destroy theses complex soil grains, as, for instance, plowing the soils when it is too wet, weakens the crop-producing power of the soil. this complexity of structure is one of the chief reasons for the difficulty of understanding clearly the physical laws governing soils.

Pore space –of soils

It follows from this description of soil structure that the soil grains do not fill the whole of the soil space. The tendency is rather to form clusters of soil grains which, though touching at many points, leave comparatively large empty spaces. This pore-space in soils varies greatly, but with a maximum of about 55 per cent. In soils formed under arid conditions the percentage of pore-space is somewhere in the neighborhood of 50 per cent. There are some arid soils, notable gypsum soils, the particles of which are so uniform in size that the pore-space is exceedingly small. Such soils are always difficult to prepare of agricultural purposes.

It is the pore space in soils that permits the storage of soil-moisture; and it is always important for the farmer so to maintain his soil that the pore-space is large enough to give him the best results, not only for the storage of moisture, but for the entrance into the soil fit for the habitation of plans. This can always be best accomplished, as will be shown hereafter, by keep plowing, when the sol is not too wet, the exposure of the plowed soil to the elements, the frequent cultivation of the soil through the growing season, and the admixture of organic matter. The natural soil structure at depths not reached by the plow evidently cannot be vitally changed by the farmer.

Hygroscopic soil-water

Under normal conditions, a certain amount of water is always found in all things occurring naturally, soils included. Clinging to every tree, stone or animal tissue is a small quantity of moisture varying with the temperature, the amount of water in the air, and with other well-known factors. It is impossible to rid any natural substances wholly of water without heating it to a high temperature. This water which, apparently, belongs to all natural objects din commonly called hygroscopic water. Hilgard states that the soils of the arid regions contain, under a temperature of 15° C. and an atmosphere saturated with water, approximately 5½ per cent of hygroscopic water. In fact, however, the air over the arid region is far form being saturated with water and the temperature is even higher than 15° C., and the hygroscopic moisture actually found in the soils of the dry farm territory is considerable conditions prevailing in the great Basin the hygroscopic water of soils varies from 0.75 per cent 3½ per cent; the average amount is not far form 1½ per cent.

Whether or not the hygroscopic water of soil is of value in plant growth is a disputed question. Hilgard believes that the hygroscopic moisture can be of considerable help in carrying plants through rainless summers, and further, that its presence prevents the heating of the soil particles to a point dangerous to plant roots. Other authorities maintain earnestly that the hygroscopic soil-water is practically useless to plants. Considering the fact that wilting occurs ling before the hygroscopic water contained in the soil is reached, it is very unlikely that water so held is of any real benefit to plant growth.

Gravitational water

It often happens that a portion of the water in the soil is under the immediate influence of gravitation. For instance, a stone which normally, is covered with hygroscopic water is dipped into water. The hygroscopic water is not thereby affected, but as the stone is drawn out of the water a good part of the water runs off. This is gravitational water. That is, the gravitational water of soils is the great portion of the soil water which, filling the soil pres, flow downward through the soil under the influence of gravity. When the soil pores are completely filled, the maximum amount of gravitational water is found there. In ordinary dry-farm soils this total water capacity is between 35 and 40 per cent of the dry weight of soil.

The gravitational soil-water cannot long remain in that condition; of necessarily, the pull of gravity moves into downward, through the soil pores and if conditions are favorable, it finally reaches the standing water-table, whence it is carried to the great rivers, and finally to the ocean. In humid soils, under a large precipitation, gravitational water moves down to the standing water table after every rain. In dry-farm soils the gravitational water seldom reaches the standing water table; for, as it moves downward, it wets the soil grains and remains in the soil capillary condition as a thin film around the soil grains.

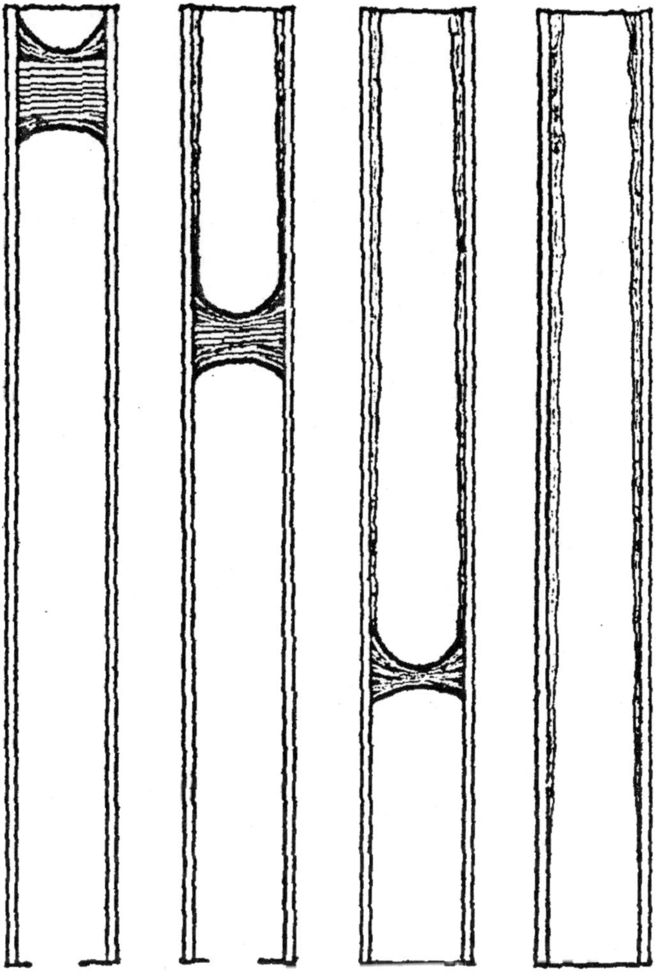

Fig. 19: Water moving downward in small tubes gradually becomes distributed over the walls of the tubes as a capillary film.

To the dry-farmers, the full water capacity is of importance only as it pertains to the upper foot of soil. if, by proper plowing and cultivation, the upper soil be loose and porous, the precipitation is allowed to soak quickly into the soil, away from the action of the wind and sun, from this temporary reservoir, the water m, in obedience to the pull of gravity, will move slowly downward to the greater soil depths, where it will be stored permanently untie needed by plants. It is for this season that dry-farmers find it profitable to plow in the fall, as soon as possible after harvesting. In fact, Campbell advocates that the harvester be followed immediately by the disk, later to be followed by the plow. The essential thing is to kept the topsoil open to a rain.

Capillary soil-water

The so called capillary soil-water is of greatest importance to the dry farmer. This is the water that clings as a film around a marble that has been dipped into water. There was natural attraction between water and nearly all known substances, as is witnessed by the fact that nearly all things may be moistened. The water is held around the marble because the attraction between the marble and the water is greater the attraction, the thicker the film; the smaller the attraction, the thinner the film will be. The water that rises in a capillary glass tube when placed in water does so by virtue of the force that makes capillary water possible is called surface tension (Fig 19).

Whenever there is a sufficient amount of water available, a thin film of water is found around every soil grain; and where the soil grain touch, or where they are very near together, water is held pretty much as in capillary tubes. No only are the soil particles enveloped by such a film, but the plant roots foraging in the soil are likewise covered; that is the whole system of soil grains and roots is covered, under favorable conditions, with a thin film of capillary water. It is the water in this form upon which plants draw during their periods of growth. The hygroscopic water and the gravitational water are of comparatively little value in plant growth.

Field capacity of soils for capillary water

The tremendously large number soil grains found in even a small amount of soil makes it possible for the soil to hold very large quantities of capillary water. To illustrate; In one cubic inch of sand soil the total surface e exposed by the soil grains varies from 42 square inches to 27 square feet; in one cubic inch of silt soil from 27 square feet to 72 square feet, and in one cubic inch of an ordinary soil the total surface exposed by the soil grains is about 25 square feet. This means that the total surface of the soil grains contained in a column of soil 1 square foot at the to sand 10 feet deep is approximately 10 acres. When even a thin film of water is spread over such a large area, it is clear that the total amount of water involved must be large. It is to be noticed, therefore, that the fineness of the soil particles previously discussed has a direct bearing upon the amount of water that soils may retain for the use of plant growth. As the fineness of the soil grains increases, the total surface increases, and the water-holding capacity also increases

Naturally, the thickness of a water film held around the soil grain is very minute. King has calculated that a film 275 millionths of an inch thick, clinging around the soil particles, is equivalent to 14.24 per cent of water in heavy clay; 7.2 per cent in a loam; 5.21 per cent in a sandy loam, and 1.41 per cent in a sandy soil.

It is important to know the largest amount of water that soils can hold in a capillary condit9ion, for upon it depend, in a measure, the possibilities of crop production under dry-farming conditions. King states that the largest amount of capillary water that can be held in sandy loams varies form 17.65 per cent to 1.67 per cent; in clay loams form 22.67 per cent to 18.16 per cent, and in humus soils (which are practically unknown in dry-farm sections) from 44.72 per cent o 21.29 per cent. These results were not obtained under dry-farm conditions and; must be confirmed by investigation of arid soils.

The water that fall upon dry-farms is very seldom sufficient in a quantity to reach the standing water table, and it is necessary, therefore, to determine the largest percentage of water that a soil can hold under influence of gravity down to a depth of 8 or 10 feet – the depth to which the roots penetrate and in which root action is distinctly felt. This is somewhat difficult to determine because the many conflicting factors acting upon the soil water are seldom in equilibrium. Moreover, a considerable time must usually elapse before the rain water is thoroughly distributed throughout the soil. For instance, in sandy soils, the downward descent of water is very rapid; in clay soils here the preponderance of fine particle makes minute soils pores, there is considerable hindrance to the descent of water, and it may take weeks or month for equilibrium to be established. It is believed that in dry-farm district, where the major part of the precipitation comes during wither, the early springtime before the spring rains come, is the best time for determining the maximum water capacity of a soil. at that season the water capacity of as soil at that season the water dissipating influences, such as sunshine and high temperature, are at a minimum, and a sufficient time has elapsed to permit the rains of fall and wither to distribute themselves uniformly throughout the soil. in districts of high summer precipitation, the alter fall after a fallow season will probably be the best time of the determination of the fields water capacity (Fig. 20)

Experiments on this subjects have been conducted at the Utah station. As a result of several thousand trials its was found that, in the spring, a uniform, sandy loam soil of true arid properties contained form year to year, and average of nearly 16½ per cent of water to a depth of 8 feet. This appeared to be practically the maximum water capacity of that soil under field conditions and it may be called the field capacity of that soil for capillary water. Other experiments of dry farms showed the field capacity of clay soil to a depth of 8 feet to 19 per cent; of a clay loam, to be 18 per cent; of a loam, 17 per cent; of another loam somewhat more sandy, 16 per cent; of a sandy loam 14½ per cent, and of a very sandy loam, 14 per cent. leather found that in the calcareous arid soil of India the upper 5 feet contained 18 percent of water at the close of the wet season.

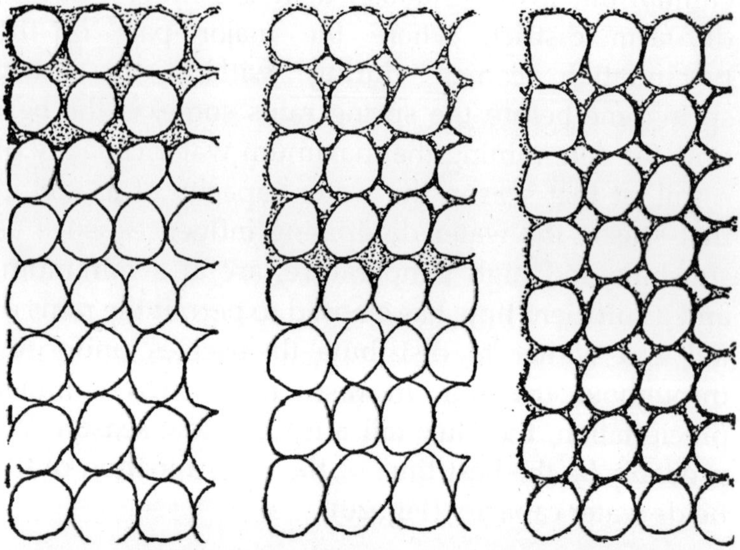

Fig. 20: Rainwater moving downward through soil becomes changed into a capillary film of water around the soil particles.

It may be concluded, therefore, that the field water capacities of ordinary dry-farm soils are not very high, ranging form 15 to 20 percent, with an average for ordinary dry-farm soils in the neighborhood of 16 or 17 per cent. Expressed another way this means that a layer of water from 2 to 3 inches deep can be stored in the soil to a depth of 12 inches. Sandy soils will hold less water than clayey ones. It must not be forgotten that in the dry-farm region are numerous types of soils among them some consisting chiefly of very fine soil grains and which would, consequently possess field water capacities above the average here stated. The first endeavor of the dry farmer should be have the soil filled to soils full field water capacity before a crop is planted.

Downward movement of soil-moisture

One of the considerations in a discussion of the storing of water in soils is the depth to which water may move under ordinary dry-farm conditions. In humid regions, where the water table is near the surface and where the rainfall is very abundant, no question has been raised concerning the possibility of the descent of water through these oils to the standing water, considerable objection, however, has been offered to the doctrine that the rainfall of arid districts penetrates the soil to any great extent. Numerous writers on the subject intimate that the rainfall under dry-farm conditions reaches at the best the upper 3 or 4 feet of soil this cannot be true for the deep rich soils of the arid region which never have been disturbed by the husbandman, are moist to very great depths. In the deserts of the Great Basin, where vegetation is very scanty, soil borings made almost anywhere will reveal the fact that moisture exist sin considerable quantities to the full depth of the ordinary soil auger, usually 10 feet. The same is true for practically every district of the arid region.

Such water has not come from below; for in the majority of cases the standing water is 50 to 500 feet below the surface. Whitney made this observation many years ago and reported it as a striking feature of agriculture in arid regions worthy of serous consideration. Investigations made at the Utah Station have shown that undisturbed soils within the Great Basin frequently contain, to a depth of 10 feet, an amount of water equivalent to 2 or 3 year of the rainfall which a normally occurs in that locality. Theses quantities of water could not be found in such soil, unless under arid conditions, water has the power to move downward to considerably greater depths than in is usually believed by dry farmers.

In a series of irrigation experiments conducted at the Utah Station it was demonstrated that on a loam soil, within a few hours after and irrigation, some of the water applied had reached the eighth foot, or at least had increased the percentage of water in the eighth foot. The following statement from these experiments shows the increase in each foot about eighteen hours after a small and also a larger irrigation.

It will be seen that in the soil that was already filled with water, the addition of water was felt distinctly to the full depth of 8 feet. Moreover, it was observed in these experiments that even very small rains caused moisture changes to considerable depths a few hours after the rain was over. For instance, 0.14 of an inch of a rainfall was felt of a depth of 2 feet with in 3 hours; 0.93 of an inch was felt of a depth of 3 feet within the same period.

To determine whether or not the natural winter precipitation, upon which the crops of a large portion of the dry-farm territory depend, penetrates the soil to any great depth a series of tests were under taken.

TABLE: 7

Water applied in inches	Time of sampling	Percentage of water in soil (foot sections)								
		1	2	3	4	5	6	7	8	Average
2.5	Before irrigation	9.57	10.55	11.78	12.92	11.92	11.41	11.75	11.49	11.43
	After irrigation	19.24	13.70	13.10	13.84	12.66	12.72	12.31	12.70	13.67
	Increase	**9.67**	**3.15**	**1.39**	**0.87**	**0.74**	**0.31**	**0.56**	**1.21**	**2.24**
7.5	Before irrigation	10.62	12.44	14.44	15.11	14.20	13.40	13.13	13.27	13.33
	After irrigation	28.83	21.83	2.05	17.40	15.87	14.66	14.21	14.15	17.75
	Increase	**13.21**	**9.39**	**5.61**	**2.29**	**1.67**	**1.26**	**1.08**	**0.88**	**4.42**

TABLE: 8

Date	Percentage Of Water In Each Foot Of Soil								
	1	2	3	4	5	6	7	8	Average
Sept. 8, 1902	6.37	7.32	8.17	8.55	8.26	9.29	1010	10.38	8.56
April 24, 1903	19.29	19.08	18.83	16.99	13.61	12.62	12.24	12.37	15.63
Increase	**12.92**	**11.76**	**10.66**	**8.44**	**5.35**	**3.33**	**2.14**	**1.99**	**7.07**
Aug. 24, 1906	8.33	7.63	8.42	9.66	11.30	10.75	9.59	7.93	9.20
May 1, 1907	18.17	16.73	17.96	16.88	16.59	16.25	14.98	13.48	16.38
Increase	**9.84**	**9.10**	**9.54**	**7.22**	**5.29**	**5.50**	**5.39**	**5.55**	**7.18**

At the close of the harvest in August or September the soil was carefully sampled to a depth of 8 feet, and in the following spring similar samples were taken on the same soils to the same depth. In every case it was found that the winter precipitation had caused moisture changes to the dull depth reached by the soil auger. Moreover, these changes were so great as to lead the investigators to believe that moisture changes had occurred to greater depths. The following shows some of the results obtained.

In districts where the major part of the precipitation occurs during the summer the same law is undoubtedly in operation; but, since evaporation is most active in the summer it is probable that a smaller proportion reaches the greater soil depths. In the Great Plains district, therefore, greater care will have to be exercised during the summer in securing proper water storage that in the Great Basin, for instance. The principle is, nevertheless, the same. Burr, working under Great Plains conditions in Nebraska, has shown that the spring summer rains penetrate the soil to the depth of 6 feet, the average depth of the borings, and that in undoubtedly affects the soil moisture to the depth of 10 feet. In general, the dry-farmer may safely accept the doctrine that the water that falls upon his land penetrates the soil far beyond the immediate reach of the sun, though no so far away that plants roots cannot make use of it.

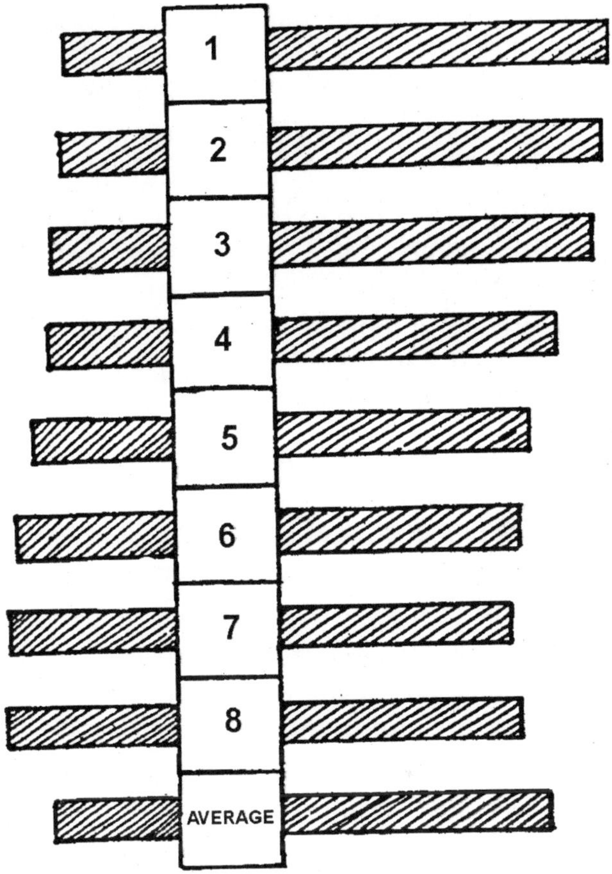

Fig. 21: Diagram to illustrate the degree and depth to which the precipitation of fall, winter, and earliest spring is found in the soil at seed time. Lines on the left indicate the percentage of water in the soil in the fall; those on the right, the percentage of water in the soil in the spring at seed time.

Importance of a moist subsoil

In the consideration of the down ward movement of soil water it is to be noted that it is only when the soil is tolerably moist that the natural precipitation moves rapidly and freely to the deeper soil layers. When the soil is dry, the downward movement of water is much slower and the bulk of the water is them stored near the surface where the loss of moisture goes on most rapidly. It has been observed repeatedly in the investigation at the Utah Station that when properly cultivated, the precipitation penetrates farther and farther into the soil with every year of cultivation. For example, on dry-farm, the soil of which is clay loam,, and which was plowed in the fall of 19.4 and farmed annually there after, the eight foot contained in the spring of 19.5, 6.59 per cent of moisture; in the spring of 19.6, 13.11 per cent, and in the spring o 19.7, 14.75 per cent of moisture. On another farm, within a very sandy soil and also plowed in the fall o9f 1904, there was found in the eighth foot in the spring o 1905 5.63 percent of moisture in the spring of 1906, 11.41 per cento of moisture, and in the spring of 1907, 15.49 per cent of moisture. In both of these typical cases it is evident that as the topsoil was loosened, the full field water capacity of the soil was more nearly approached to a greater depth. It soil seem that, as the lower soil layers are moistened, the water is enabled, so to speak, to slide down more easily into the depths of the oil.

This is a very important principle for the dry-farmers to understand. It is always dangerous to permit the soil of dry farm to become very dry, especially below the first foot. Dry farms should be so manipulated that even at the harvesting season a comparatively large quantity of water remains in the soil to a depth of 8 feet or more. The larger the quantity of water in the soil in the fall, the more readily and quickly will the water that falls on the land during the resting period of fall, winter, and early spring sing into the soil and move away form the topsoil. The top or fist foot will always contain the largest percentage of water because it is the chief receptacles o the water that falls as rain or snow, but when the subsoil is properly moist, the water will more completely leave the topsoil. Further, crops planted on a soil saturated with water to depth of 8 feet are almost certain to mature and yield well.

If the field-water capacity has not been filled, there is always the danger that an unusually dry season or a series of hot winds or other like circumstances may either seriously injure the cop or cause a complete failure. The dry-farmers should keep a surplus of moisture in the soil to be carried over form year to year, just as the wise business man maintain a sufficient working capital for the needs of his business. In fact, it is often safe to advise the prospective dry-farmer to plow his newly cleared or broken land carefully and then to grow no crop on it the first year, so that, when crop production begin, the soil will have stored in it an amount of water sufficient to carry a crop over a period of drought. Especially in districts of very low rainfall is this practice to be recommended. In the Great Plants area, where the summer rains tempt the farmer to give less attention to the soil moisture problem that in the dry districts with winter precipitation, farther West, it is important that a fallow season be occasionally given the land to prevent the store of soil moisture form becoming dangerously low.

To what extent is the rainfall stored in soils?

What proportion of the actual amount of water falling upon the soil can be stored in the soil and carried over from season to season? This question naturally arises in view of the conclusion that water penetrates the soil to considerable depths. There is comparatively little available information with which to answer this question, because the great majority of student of soil moisture have concerned themselves wholly with the upper two, three, or four feet of soil the results of such investigation are practically useless in answering this question. In humid regions it may be very satisfactory to confine soil-moisture investigations to the upper few feet; but in arid regions, where dry farming is a living question, such a method leads to erroneous or incomplete conclusions.

Since the average field capacity of soils for water is about 2.5 inches per foot, it follows that it is possible to store 25 inches or water in 10 feet of soil. This is form two to one and a half times one year's rainfall over the better dry-farming sections. The oretically, therefore there is no reason why the rainfall of one season or more could not be stored in the soil. careful investigations have borne out this theory. Atkinson found, for example, at the Montana Station, that soil which to a depth of 9 feet contained 7.7 per cent of moisture in the fall contained 11.5 per cent in the spring and after carrying it through the summer by proper methods of cultivation, 11 per cent.

It may certainly be concluded form this experiment that it is possible to carry over the soil moisture from season to season. The elaborate in investigations sat the Utah Station have demonstrated that the winter precipitation, that is, the precipitation that comes during the wettest period of the year, may be retained in a large measure in the soil. Naturally, the amount of the natural precipitation accounted for in the upper eight feet will depend upon the dryness of the sol at the time the investigation commenced. It at the beginning of the wet season the upper eight feet of soil are fairly well stored with moisture, the precipitation will move down to even greater depths, beyond the reach of the soil auger. If, on the other had, the soil is comparatively dry at the beginning will distribute itself through the upper few feet, and thus be readily measured by soil auger.

In the Utah investigations it was found that of the water which fell as rain and snow during the winter, as high as 95½ per cent was found stored in the first eight feet of soil at the beginning of the growing season. Naturally, much smaller percentages were also found, but on an average, in soils somewhat dry at the beginning of the dry season, more than three fourths of the natural precipitation was found stored in the soil in the spring. The following table shows some of these summary results:-

TABLE: 9
PROPORTION OF RAINFALL STORED IN THE SOIL

Period	Percent of water in soil in fall (Depth of 8 ft.)	Rainfall during period (Inches)	Percent of precipitation found in the spring (to a depth of 8 ft.)	Soil
Sept. 12, 1902 – April 16, 1903	8.78	8.51	87.59	Sandy Loam
Aug. 23, 1904 – April 22, 1905	7.87	7.94	95.56	Sandy Loam
Sept. 8, 1905 – April 28, 1906	8.83	12.14	82.61	Sandy Loam
Oct. 8, 1906 – April 29, 1907	9.10	16.17	62.77	Sandy Loam
Sept. 14, 1907 – April 23, 1908	11.03	3.38	67.55	Sandy Loam
July 27, 1904 – April 15, 1905	12.34	10.51	93.17	Clay
Aug. 8, 1904 – April 5, 1905	7.73	7.27	64.80	Sand
July 28, 1905 –May 7, 1906	11.04	10.65	81.13	Loam

While the results exhibited in the above table were all obtained in a locality where the bulk of the precipitation comes in the winter, yet similar results would undoubtedly obtained where the precipitation occurs mainly in the summer. The storage of water in the soil cannot be a whit less important on the Great Plains than in the great Basin. In fact, Burr has clearly demonstrated for western Nebraska that over 50 per cent of the rainfall of the spring and summer may be stored in the soil to the depth of six feet. Without question, some is stores also at greater depths.

All the evidence at had shows that a large portion of the precipitation falling upon properly prepared soil, whether it is in summer or winter, is stored in the soil until evaporation is allowed to withdraw it. Whether or not water so stored may be made to remain in the soil throughout the season or the year will be discussed in the next chapter. It must be said, however, that the possibility of storing water in the soil, that is, making the water descend to relatively great soil depths away from the immediate and direct action of the sunshine and winds, is the most fundamental principle in successful dry-farming.

The fallow

It may be safely concluded that a large portion of the war falls as rain or snow may be stored in the soil to considerable depth (eight feet or more). However, the question remains, is it possible to store the rainfall of successive years in the soil for the use of one crop? In short, Does the practice of clean fallowing or resting the ground with proper cultivation for one season enable the farmer to store in the soil the larger portion of the rainfall of two years, to be used of one crop? It is unquestionably true, as will be shown later, that clean fallowing or "summer tillage" is one of the oldest and safest practices of dry farming as practiced in the West, but it is no generally understood why fallowing is desirable.

Considerable doubt has recently been cast upon the doctrine that one of the beneficial effects of fallowing in dry-farming is to store the rainfall of successive season in the soil for the use of one corp. since it has been shown that a large proportion of the winter precipitation can be stored in the soil during the wet season, it merely becomes a question of the possibility of preventing the evaporation of this water during the drier season. As will be shown in the next chapter, his can well be affected by proper cultivation.

There is no good reason therefore, for believing that he precipitation of successive season may not be added to water already stored in the soil. King has hown that fallowing the soil one year carried over per square foot, in the upper four feet, 9.38 pounds of water more that was found in a cropped soil in a parallel experiments; and, moreover, the beneficial ffects of this water advantage was felt for a whole ucceeding season. King concludes, therefore, that ne of the advantages of fallowing is to increase the noisture content of the soil. The Utah experiment that ie tendency of fallowing is always to increase the oil moisture content. In dry-farming, water is the critical factor and any practice that helps to conserve the water should be adopted. For that reason, fallowing which gathers soil moisture, should be strongly advocated. In chapter IX another important value of the fallow will be discussed.

In view of the discussion in this chapter it is easily understood why students of soil-moistures have not found a material increase in soil moisture due to fallowing. Usually such investigations have been made to shallow depths which already were fairly well filled with moisture. Water falling upon such soils would find beyond the depth reached by the soil augers, and it became impossible to judge accurately of the moisture-storing advantage of the fallow. A critical analysis of the literature on this subject will reveal the weakness of most experiments in this respect.

It may be mentioned here that the only fallow that should be practiced by the dry-farmer is the clean fallow. Water storage is manifestly impossible when crops are growing upon a soil. a healthy crop of sagebrush, sunflowers, or other weeds consumes as much water as first-class stand of corn wheat or potatoes. Weeds should be abhorred by the farmer. A weedy fallow is a sure forerunner of a crop failure. How to maintain a good fallow is discussed in Chapter VIII, under the head of Cultivation. Moreover, the practice of fallowing should be varied with the climatic conditions. In district of low rainfall 10-15 inches, the land should be clean summer fallowed every other year; under very low rainfall perhaps even two out of three years; in district of more abundant rainfall, 15-20 inches, perhaps one year out of every three of four is sufficient. Where the precipitation comes during the growing season, as in the Great Plains area fallowing for the storage of water is less important than where the major part of the rainfall comes during the fall and wither. However, any system of dry farming that omits fallowing wholly form its practices ins in danger of failure in dry years.

Deep lowing for water storage

It has been attempted in this chapter to demonstrate that water falling upon a soil may descend to great depths, and may be stored in the soil from year to year to year, subject to the needs of the crop that may be planted. By what cultural treatment may this downward descent of the water accelerated by the farmer? First and foremost by plowing at the right depth. Plowing at the right time and to the right depth. Plowing should be down deeply and thoroughly so that the falling water may immediately be drawn down to the full depth of the loose, spongy, plowed soil, away form the action of the sunshine or winds. The moisture thus caught will slowly work its way down into the lower layer of the soil. Deep plowing is always to be recommended for successful dry-farming.

In humid districts where there is a great difference between the soil and the subsoil, it is often dangerous to turn up the soil and the subsoil. But in arid districts where there is no real differentiation between the soil and the subsoil, deep plowing may safely be recommended. True, occasionally, soils are found in the dry-farm territory, which are under laid near the surface by and inert clay or infertile layer of lime or gypsum, which forbids the farmer putting the plow too deeply into the soil. Such soils, however, are seldom worth while trying for dry farm purposes. Deep plowing must be practiced of the best dry farming results.

It naturally follows that sub soiling should be a beneficial practice on dry –farm. Whether or not the great cost of sub soiling is offset by the resulting increased yields is an open question: it is, in fact, quite doubtful. Deep plowing down at the right time and frequently enough is possible sufficient. By deep plowing is meant stirring or turning the soil to a depth of six to ten inches, below the surface of the land.

Fall plowing for water storage

It is not alone sufficient to plow and to plow deeply; it is also necessary that the plowing be done at the right time. In the very great majority of cases over the whole dry-farm territory, plowing should be done in the fall. There are three reasons for this: First, after the crop is harvested, the soil should be stirred immediately, so that it can be exposed to the full action of the weathering agencies, whether the winters be open or closed. If from any reason plowing cannot be done early it is often advantageous to follow the harvester with as disk and to plow the later when convenient. The chemical effect on the soil resulting from the weathering, made possible by a fall plowing, as will be shown in chapter IX, is of itself so great as to warrant the teaching of the general practice of fall plowing. Secondly, the early stirring of the soil prevents evaporation of the moisture in the soil during the late summer and the fall. Thirdly, in the parts of the dry-farm territory where much precipitation occurs in the fall winter, or early spring, fall plowing permits much of this precipitation to enter the soil and be stored there till needed by plants.

A number of experiments stations have compared plowing down in the early fall with plowing done late in the fall or in the spring, and with almost no exception it has been found that early fall plowing is water conserving and in other ways advantageous. It was observed on a Utah dry-farm that the fall plowed land contained, to a depth of 10 feet, 7.47 acre-inches more water than the adjoining spring plowed land – a saving of nearly one half of a year's precipitation. The ground should be plowed in the early fall as soon as possible after the crop is harvested. It should then be left in the rough throughout the winter, so that it may be mellowed an broken down by the elements. The rough land further has a tendency to catch and hold the snow that may be blown by the wind, thus insuring a more even distribution of the water from the melting snow.

A common objection to fall plowing is that the ground is so dry in the fall that it does not plow up well, and that the great, dry clods of earth do much to injure the physical condition of the soil. It is very doubtful if such an objection is generally valid, especially if the soil is so cropped as to leave a fair margin of moisture in the soil at harvest time. The atmospheric agencies will usually break down the clods, and the physical result of the treatment will be beneficial. Undoubtedly, the fall plowing of dry land is somewhat difficult, but the good results more than pay the farmer for his trouble. Late fall plowing after the fall rains and have softened the land ins preferable to spring plowing. If for any reason the farmer feels that he must practice spring plowing, he should do it as early as possible in the sprig. Of course, it is inadvisable to plow the soil when it is so wet as to injure its tilth seriously, but as soon as that danger period has passed, the plow should be placed in the ground. The moisture in the soil will thereby be conserved, and whatever water may fall during the spring months will be conserved also. This is of especial importance in the Great Plains region and in any district where the precipitation comes in the spring and winter months.

Likewise, after fall plowing, the land must be well stirred in the early spring with the dist harrow or a similar implement, to enable the spring rains to enter the soils easily and to prevent the evaporation of the water already stored. Where the rainfall is quite abundant and the plowed land has been beaten down by the frequent rains, the land should be plowed again in the spring. Where such conditions do not exist, the treatment of the soil with the disk and harrow in the spring is usually sufficient.

In recent dry-farm experience it has been fairly completely demonstrated that, providing the soil is well stored with water, crops will mature even if no rain falls during the growing season. Naturally, under most circumstances, any rains that may fall on a well prepared soil during the season of crop growth will tend to increase the crop yield, but some profitable yield is assured in spite of the season, if the soil is well stored with water at seed time. This is an important principle in the system of dry – farming.

CHAPTER VIII

REGULATING THE EVAPORTAION

The demonstration in the last chapter that the water which falls as rain or snow may be stored in the soil for the use of plants is of first importance in dry-farming, for it makes the farmer independent, in large measure, of the distribution of the rainfall. The dry-farmer who goes into the summer with a soil well stored with water cares little whether summer rains come or not, for he knows that his crops will mature in spite of external drought. In fact, as will be shown later, in many dry-farm sections where the summer rains are light they are a positive detriment t the farmer who by careful farming has stored his deep soil with an abundance of water. Storing the soil with water, is however, only the first step in making the rains of fall, wither, or the preceding year available of plant growth, as soon as warm growing weather comes, water dissipating forces come into play, and water is lost by evaporation. The farmer must, therefore, use all precautions to keep the moisture in the soil until such time as the roots of the crop may draw it into the plants to be used in plant production.

This is, as far as possible, direct evaporation of water form the soil must be prevented.

Few farmers rally realize the immense possible annual evaporation in the dry farm territory. It is always much larger than the total annual rainfall. In fact, an arid region may be defined as one in which under natural conditions several times more water evaporates annually from a free water surface that falls as rain and snow. For that reason many students of aridity pay little attention to temperature, relative humidity, or winds, and simply measures the evaporation form a free water surface in the locality in question. In order to obtain a measure of the aridity, MacDougal has constructed the following table, showing the annual precipitation and the annual evaporation at several well known localities in the dry farm territory.

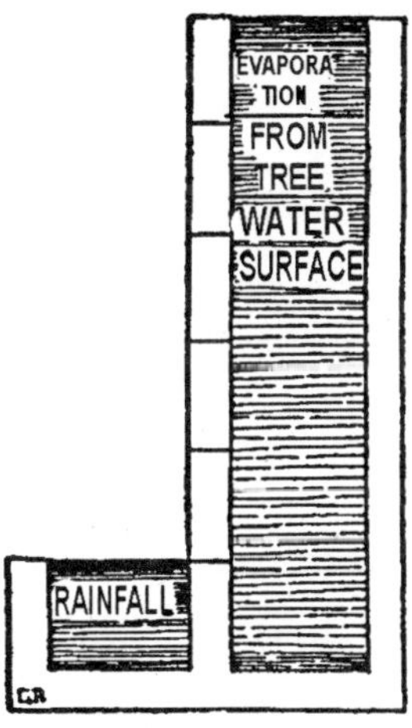

Fig. 22: Annual rainfall and evaporation in arid region compared. The high evaporation rate makes necessary thorough farming.

True, the localities included in the following table are extreme, but they illustrate the large possible evaporation, ranging from about six to thirty-five time the precipitation. (See Fig 22) At the same time it must be borne in mind that while such rates of evaporation may occur from free water surfaces the evaporation from agricultural soils under like condition sis very much smaller.

TABLE: 10

Place	Annual precipitation (Inches)	Annual Evaporation (Inches)	Ratio
El Paso, Texas	9.23	80	8.7
Fort Wingate, New Mexico	14.00	80	5.7
Fort Yuma, Arizona	2.84	100	35.2
Phoenix, Arizona	7.06	90	12.7
Tucson, Arizona	11.74	90	7.7
Mohave, California	4.97	95	19.1
Hawthorne, Nevada	4.50	80	17.5
Winnemucca, Nevada	8.51	80	9.6
St. George, Utah	6.46	90	13.9
Fort Duchesne, Utah	6.49	75	11.6
Pineville, Oregon	9.01	70	7.8
Lost River, Idaho	8.47	70	8.3
Laramie, Wyoming	9.81	70	7.1
Torres Mexico	16.97	100	6.0

To understand the methods employed for checking evaporation from the soil, it is necessary to reviews briefly the conditions that determine the evaporation of water into the air, and the manner in which water moves in the soil.

The formation of water vapor

Whenever is left freely exposed to their air, it evaporates; that is, it passes into the gaseous state and mixes with the gases of the air. Even snow and ice give off water vapor, though in very small quantities. The quantity of water vapor which can enter a given volume of air is definitely limited. For instance, at the temperature of freezing water 2.216 grains of water vapor can enter one cubic foot of air, but no more. When air contains all the water possible it is said to be saturated, and evaporation then ceases. The practical effect of this is the well known experience that on the seashore, where the air is often very nearly fully saturated with water vapor, the drying of clothes goes on very slowly, where as in the interior, like the dry-farming territory away form the ocean where the air is far from being saturated, drying goes on very rapidly.

The amount of water necessary to saturate air varies greatly with the temperature, as may be seen from the Table 11.

It is to be noted that the temperature increases, the amount of water that may be held by the air also increases; and proportionately more rapidly than the increase in temperature. This generally well understood in common experience, is in drying clothes rapidly by hanging them before a hot fire. At a temperature of 100°F., which is often reached in portions of the dry-farm territory during the growing season, a given volume of air can hold more than nine times as much water vapor as at the temperature of freezing water. This is an exceedingly important principle in dry farm practices for it explains the relative easy possibility of storing water during the fall and winter when the temperature is low and the moisture usually abundant, and the greater difficulty of storing the rain that falls largely as in the Great Plains area, in the summer, when water dissipating forces are very active. This law also emphasizes the truth that it is in times of warm weather that every precaution must be taken to prevent the evaporation of water form the soil surface.

TABLE: 11

Temperature Fahrenheit (Degrees)	Weight of Water Vapor that can be held in one cubic foot of air (In Grains)	Difference
0	0.545	
32	2.126	
40	2.862	0.736
50	4.089	1.227
60	5.756	1.667
70	7.992	2.236
80	10.949	2.957
90	14.810	3.861
100	19.790	4.980

It is of course well understood that the atmosphere as a whole is never saturated with water vapor. Such saturation is at the best only local, as, for instance, on the seashore during quiet days when the layer of air over the water may be fully saturated, or in a field containing much water from with, on quiet warm days, enough water may evaporate to saturate the layer of air immediately upon the soil and around the plants. Whenever, in such cases, the air begins to move and the wind blows, the saturated air is mixed with the larger portion of unsaturated air, and evaporation is again increased. Meanwhile, it must be borne in mind that into a layer of saturated air resting upon a field of growing plants very little water evaporates, and that the chief water dissipating power of winds lies in the removal of this saturated layer. Winds or air movements of any kind therefore, become enemies of the farmers who depends upon a limited rainfall.

The amount of water actually found in a given volume of air at a certain temperature, compared with the largest amount it can hold, is called, is called the relative humidity of the air. As shown in Chapter IV, the relative humidity becomes smaller as the rainfall decreases. The lower the relative humidity is at a given temperature, the more rapidly will water evaporate into the air. There is no more striking confirmation of this law than the fact that at a temperature of 90° sunstrokes an similar ailments are reported in great number from New York, while the people of Salt Lake City are perfectly comfortable. In New York the relative humidity in summer is about 73 percent; in Salt Lake City, about 35 per cent. At a high summer temperature evaporation from the skin goes on slowly in New York and rapidly in Salt Lake City, with the resulting discomfort. Similarly, evaporation form soils goes on rapidly under a low and slowly under a high percentage of relative humidity.

Evaporation from water surfaces is hastened, therefore, by:

(1) an increase in the temperature,

(2) an increase in the air movements or winds, and

(3) a decrease in the relative humidity.

The temperature is higher; the relative humidity lower, and the winds usually more abundant in arid than in humid regions. The dry farmer must subsequently use all possible precautions to prevent evaporation from the soil.

Conditions of evaporation from soils

Evaporation does not alone occur from a surface of free water. All wet or moist substances lose by evaporation most of the water that they hold, providing the conditions of temperature and relative humidity are favorable. Thus from a wet soil, evaporation is continually removing water. Yet, under ordinary conditions, it is impossible to remove all the water, for a small quantity is attracted so strongly by the soil particles that only a temperature above the boiling point of water will drive it out. This part of the soil is the hygroscopic moisture spoken in the last chapter.

Moreover, it must kept in mind that evaporation does not occur as rapidly from wet soils as from a water surface, unless all the soil pores are so completely filled with water that the soil surface is practically a water surface. The reason for this reduced evaporation from a wet soil is almost self-evident. There is a comparatively strong attraction between soil and water, which enables the moisture to cling as a thin capillary film around the soil particles, against the force f gravity. Ordinarily, only capillary water is found in well – tiled soil, and the force causing evaporation must be strong enough to overcome this attraction besides changing the water into vapor.

The less water there is in a soil, the thinner the water film, and the more firmly is the water held. Hence, the rate of evaporation decreases with the decrease in the soil moisture. This law is confirmed by actual field tests. For instance, as an average of 274 trials made at the Utah Station, it was found that three soils, otherwise alike, that contained, respectively, 22.63 per cent 17.14 per cent, and 12.75 per cent of water lost in two weeks, to a depth of eight feet, respectively 21.0, 17.1, and 10.0 pounds of water per square foot. Similar experiments conducted elsewhere also furnish proof to the correctness of this principle. From this point of view the dry-farmer does not want his soils to be unnecessarily moist. The dry – farmers, can reduce the per cent of water in the soil without diminishing the total amount of water by so treating the soil that the water will distribute itself considerable depths. This brings into prominence again the practices of fall plowing, deep plowing, sub-soiling, and the choice of deep soils for dry – farming.

Very much for the same reasons, evaporation goes on more slowly form water in which salt or other substances have been dissolved. The attraction between the water and the dissolved salt seems to be strong enough to resist partially the force causing evaporation. Soil-water always contains some of the soil ingredients in solute on, and consequently under the given condition evaporations occurs more slowly from soil water than from pure water. Now the more fertile a soil is, that is, the more soluble plant-food it contains, the more material will be dissolved in the soil water, and as a result the more slowly will evaporation. While these conditions may have little value in the eyes of the farmer who is under and abundant rainfall, they are of great importance to the dry-farmer. It is only by utilizing every possibility of conserving water and fertility that dry farming may be made a perfectly safe practices.

Loss by evaporation chiefly at the surface

Evaporation goes on from every wet substance. Water evaporates therefore from the wet soil grains under surface as well as from those at the surface. In developing a system of practice, which will reduce evaporation to a minimum it must be learned whether the water which evaporates from the soil particles faro below the surface is carried in large quantities into the atmosphere and thus lost to plant use. Over forty years ago, Nessler subjected this question to experiment and found that the kiss by evaporation occurs almost wholly at eh soil surface, and that very little if any is lost directly by evaporation from the lower soil layers. Other experimenters have confirmed this conclusion, and very recently Buckingham, examining the same subject, fount that while there is a very slow upward movement of the soil gases into the atmosphere, the total quantity of the water thus lost by direct evaporation from soil, a foot below the surface, amounted at most to one inch of rainfall in six years. This insignificant even under, the rate of loss of water by direct evaporation from the lower soil layers increases with the porosity of the soil, that is, with the space not filled with soil particles or water, fine-grained soils, therefore, lose the least water in this manner. Nevertheless, if coarse-grained soils are well filled with water. By deep fall plowing and by proper summer fallowing for the conservation of moisture, the loss of moisture by direct evaporation from the lower soil layers need not be larger than from finer grained soils.

Thus again are emphasized the principles previously laid down that, for the most successful dry-farming, the soil should always be kept well filled with moisture, even if it means that the land, after being broken, must lie fallow for one or two season, until a sufficient amount of moisture has accumulated. Further the correlative principle is emphasized that the moisture in dry-farm lands should be stored deeply, away from the immediate action of the sun's rays upon the land surface. The necessity for deep soils is thus again brought out.

The great loss of soil moisture due to an accumulation of water in the upper twelve inches is well brought out in the experiments conducted by the Utah Station. The following is selected form the numerous data on the subject. Two soils almost identical in character, contained, two soils, almost identical in character, contained respectively 17.57 per cent and 16.55 per cent of water on an average to a depth of eight feet; that is the total amount of water held by the two soils was practically identical. Owing to varying cultural treatment, the distribution of the water in the soil was not uniform; one contained 23.22 per cent and the other 16.64 per cent of water in the first twelve inches. During the first seven days the soil that contained the highest percentage of water in the first foot lost 13.30 pounds of water, while the other lost only 8.48 pounds per square foot. This great difference was due no doubt to the fact that direct evaporation takes place in considerable quantity only in the upper twelve inches of soil, where the sun's heat has a full chance to act.

Any practice which enables the rains to sink quickly to considerable depths should be adopted by the dry farmer. This is perhaps one of the great reasons for advocating the expensive but usually effective subsoil plowing on dry farms. It is a very common experience, in the arid region, that great, deep cracks form during hot weather. From the walls of these cracks evaporation goes on, as from the topsoil and the passing winds renew the air so that the evaporation may go on rapidly. (See Fig. 23). The dry farmer must go over the land as often as needs be with some implement that will destroy and fill up the cracks that may have been formed. In a field of growing crops this is often difficult to do; but it is not impossible that hand hoeing, expensive as it is, would pay well in the saving of soil moisture and the consequent increase in crop yield.

TABLE: 12

Per cent of water in	1st foot	2nd foot	3rd foot	4th foot	5th food	6th foot	7th foot	8th foot	Average
Early Spring	20.84	2.06	19.62	18.28	18.70	14.29	14.48	13.83	17.51
Midsummer	8.83	8.87	11.03	9.59	11.27	11.03	8.95	9.47	9.88

Fig. 23: Many soils check badly. The cracks cause a loss of soil moisture. Arid soils (this picture represents a heavy lay as depicted by yon and Fippin) often crack extensively. Cultivation will prevent the loss of soil-moisture.

How soil water reaches the surface

It may be accepted as an established truth that the direct evaporation of water from wet soils occurs almost wholly at the surface. Yet it is well known that evaporation from the soil surface may continue until the soil moisture to a depth of eight or ten feet or more is depleted. This is shown by the following analyses of dry farm soil in early spring and midsummer. No attempt was made to conserve the moisture in the soil:-

In this case water had undoubtedly passed by capillary movement from the depth of eight feet to a point near the surface where direct evaporation could occur. As explained in the last chapter, water which is held as a film around the soil particles is called capillary water; and it sis in the capillary form that water may be stored in dry – farm soils. Moreover, it is the capillary soil-moisture, alone which is of real value in crop production. This capillary water tends to distribute itself uniformly throughout the oils, in accordance with the prevailing conditions and forces. If no water is removed form the soil, in course of time the distribution of the soil water will be such that the thickness of the film at any pout in the sol mass is direct resultant of the various forces acting at that particular point. There will then be no appreciable movement of the soil-moisture. Such a condition is approximated in late winter of early spring before planting begins. The distribution of water under such conditions is seen in the Table 8 of the last chapter. During the greater part of the year, of there are numerous disturbing elements that normally are active, among which the three most effective are:

(1) the addition of water to the soil by rains;

(2) the evaporation of water from the topsoil due to the more active meteorological factors during spring summer, a downfall; and

(3) the abstraction of water from the soil by plant roots.

Water, entering the soil, moves downward under the influence of gravity as gravitational water, until under the attractive influence of the soils it has been converted into capillary water and adhered to the soil particles as a film. If the soil were dry, and the film therefore this, the rain a water would move downward only a short distance as gravitational water; if the soil were wet, and the film therefore thick, the water would move down to a greater distance before being exhausted. If, as is often the case in humid districts, the soil is saturated, that is the film is as thick as the particles can hold, the water would pass right through the soil and connect with the standing water below. This, of course, is seldom the case in dry farm districts. In any soil, excepting one already saturated, the addition of water will produce a thickening of the soil-water film to the full descent of the water. This immediately destroys the conditions of equilibrium formerly existing for the moisture is not now uniformly distributed. Consequently a process of redistribution begins which continues until the nearest approach to equilibrium is restored. In this process water will pass in every direction from the wet portion of the soil to the drier; it does not necessarily mean that water will actually pass form the wet portion to the drier portion; usually, at the driest point a little water is drawn from the next, and that from the next, until the redistribution is complete. The process is very much like stuffing wool into a sack, which already is loosely filled. The new wool does not reach the bottom of the sack, yet there is more wool in the bottom than there was before.

If a plant root is actively feeding some distance under the soil surface, the reverse process occurs. At the feeding point the root continually abstracts water from the soil grains and thus makes the film thinner in that locality. This causes a movement of moisture similar to eh one above described from the wetter portions of the soil to the portion being dried out by the action of the plant-root. Soil many feet or even rods distant may assist in supplying such an active root with moisture. When the thousands of tiny root sent out by each plant are recalled, it may well be understood what a confusion of pull and counter pulls upon the soil moisture exists in any cultivated soil. in fact, the soil water film may be viewed as being in state of trembling activity, ending to place itself in full equilibrium with the surrounding contending forces which, themselves. Constantly change. Were it not that the water film held closely around the soil particles in possessed of extreme mobility, it would be possible to meet the demands of the plants upon the water at comparatively great distances. Even as it is, it frequently happens that when crops are planted too thickly on dry – farms the soil moisture cannot move quickly enough to the absorbing roots to maintain plant growth, and crop failure results. Incidentally, this point to planting that shall be proportional to the moisture contained by the soil. See Chapter XI.

As the temperature rises in spring with a decrease in the relative humidity, and an increase in direct sunshine, evaporation from the soil surface increased greatly. However, as the topsoil becomes drier, that it, as the water film becomes thinner, there is an attempt at readjustment, and water moves up-ward to take the place of that lost by evaporation. As this continues throughout the season, the moisture stored eight or ten feet or more below the surface is gradually brought to the top and evaporated, and thus to plant use.

The effect of rapid top drying of soils

As the water held by soils diminishes, and the water film around the soil grains becomes thinner, the capillary movement of the soil-water is retarded. This is easily understood by recalling that the soil particles have an attraction for water, which is of definite value, and may be measured by the thickest film that may be held against gravity. When the film is thinned, it does not diminish the attraction of the soil for water; it simply results in a stronger pull upon the water and a firmer holding of the film against the surfaces of the soil grains. The move soil-water under such conditions required the expenditure of more energy than is necessary for moving water in a saturated or nearly saturated soil. Under like conditions therefore, the thinner the soil water film the more difficult will be the upward movement of the soil water and the slower the evaporation from the topsoil.

As drying goes on, a point is reached at which the capillary movement of the water wholly ceases. This is probably when little more than the hygroscopic moisture remains. In fact, very dry soil and water repel each other. This is shown in the common experience of driving along a road in summer, immediately after a light shower. The masses of dust were wetter only on the outside, an as the wheels pass through them the dry dust revealed. It is an important fact that very dry soil furnishes a very effective protection against the capillary movement of water.

In accordance with the principle above established, if the surface soil could be dried to the point where capillarity is very slow, the evaporation would be diminished o almost wholly stopped. More than a quarter of a century ago, Eser showed experimentally that soil water may be saved by drying the surface soil rapidly. Under dry-farm conditions it frequently occurs that the draft upon the water of the soil is so great that nearly all the water is quickly and so completely abstracted from the upper few inches of soil that they are left as an effective protection against further evaporation. For instance, in localities where hot dry winds are of common occurrence, the upper layer of soil is sometimes completely dried before the water in the lower layers can be slow capillary movement reach top. The dry soil layer then prevents further loss of water, and the wind because on its intensity has helped to conserve the soil moisture. Similarly in localities where the relative humidity is low, the sunshine abundant, and the temperature high, evaporation may go on so rapidly that the lower soil layers cannot supply the; demands made, and the topsoil then dried out so completely as to form a protective covering against further evaporation. It is on this principle that the native desert soils of the United States untouched by the plow, and the surfaces of which are sun-baked are often found to possess large percentages of water at lower depths. Whitney recorded this observation with considerable surprise, many years ago and other observers have found the same conditions at nearly all points of the arid region. This matter has been subjected to further study by Buckingham, who placed a variety of soils under artificially arid and humid conditions. It was found in every case that the initial evaporation was greater under arid conditions, but as the process went on and the topsoil; of the arid soil became dry, more water was lost under humid conditions. For the whole experimental period, also, more water was lost under humid conditions. It was notable that the dry protective layer was formed more slowly on alkali lands for dry farm purposes. All in all, however, it appears "that under very arid conditions a soil automatically protect itself from drying by the formation of a natural mulch on the surface."

Naturally, dry-farm soils differ greatly in their power of forming such a mulch. A heavy clay or a light sandy soil appears to have less power of such automatic protection than a loamy soil an admixture of limestone seems to favor the formation of such a natural protective mulch. Ordinarily, the farmer can further the formation of a dry topsoil layer by stirring the sol thoroughly. This assists the sunshine and the air to evaporate the water very quickly. Such cultivation is very desirable for other persons also, as will soon be discussed. Meanwhile, the water dissipating forces of the dry-farm section are not wholly objectionable, for whether the land be cultivated or not, they tend to hasten the formation of dry surface layer of soil which guard against excessive evaporation. It is in moist cloudy weather, when the drying process is slow that, that evaporation causes the greatest losses of soil-moisture.

The effect of shading

Direct sunshine is, next to temperature, the most active cause of rapid evaporation from moist soil surfaces. Whenever, therefore evaporation is not rapid enough to form a dry protective layer of top-soil, shading helps materially in reducing surface losses of soil water. Under very arid conditions, however, it is questionable whether in all cases shading has a really beneficial, effect, though under semiarid or sub-humid conditi0ons the benefits derived from shading in 1873 that the shading due to the forest cover reduced evaporation 62 per cent, and many experiments since that day have confirmed this conclusion. At the Utah Station, under arid conditions, was found that shading a pot of soil which otherwise was subjected to water dissipating influences, saved 29 per cent of the loss due to evaporation from a pot which was not shaded. This principle cannot be applied very greatly in practice, but it points to a somewhat thick planting, proportioned to the water held by the soil it also shows a possible benefit to be derived form the high header straw which is allowed to stand for several weeks in dry-farm sections where the harvest comes early and the fall plowing is done, late, as in the mountain states. The high header stubble shades the ground very thoroughly. Thus the stubble may be made to conserve the soil-moisture in dry-farm sections, where grain is harvested by the "header" method.

Fig. 24: Alfalfa in cultivated rows. This practice is employed to make possible the growth of alfalfa and other perennial crops on arid lands without irrigation.

A special case of shading is the mulching of land with straw or other barnyard litter, or with leaves, as in the forests. Such mulching reduces evaporation, but only in part, because of its shading action, since it acts also as a loose top layer of soil matter breaking communication with the lower soil layers.

Whenever the soil is carefully stirred, as will be described, the value of shading as a means of checking evaporation disappears almost entirely. It is only with soils which are tolerably most at the surface that shading acts beneficially.

The effect of tillage

Capillary soil moisture moves from particle to particle until the surface is reached. The c loser the soil grains are packed together, the greater the number of points of contact, and the more easily will the movement of the soil moisture proceed. If by any means a layer of the soil is so loosened as the reduce the number of point to contact, the movement of the soil moisture is correspondingly hindered. The process is somewhat similar to the experience in large railway stations. Just before train time a great crowd of people is gathered outside of the gates ready to show their tickets. If one gate is opened, a certain number of passengers can pass through each minute; if two are opened nearly twice as many may be admitted in the same time; if more gates are opened the passengers will be able to enter the train more rapidly. The water in the lower layers of the soil is ready to move upward whenever a call is made upon it. To reach the surface it must pass from soil grain to soil grain, and the larger the number of grains that touch, the more quickly and easily will the water reach the surface, for the points of contact of the soil particles may be likened to the gates of the railway station. Now if, by a through stirring and loosening of the topsoil, the number of points of contact between the top and subsoil is greatly reduced the upward flow of water is thereby largely checked. Such a loosening of the topsoil has come to be called cultivation, and includes plowing, harrowing, disking hoeing and other cultural operations by which the top and the subsoil is undoubtedly the main reason for the efficiency of cultivation, but it is also to be remembered that such stirring helps to dry the top soil very thoroughly, and as has been explained a layer of dry soil of itself is a very effective check upon surface evaporation. (See Fig. 25)

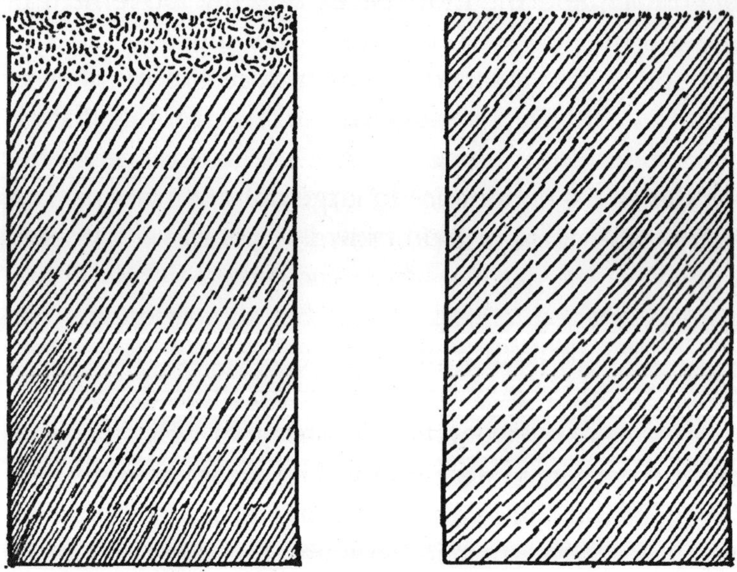

Fig. 25: Tillage forms a loose dry mulch on the land surface, which prevents evaporation.

That the stirring or cultivation of the topsoil really does diminish evaporation of water from the soil has been shown by numerous investigations. In 11868, Nessler found that during six weeks of an ordinary

German summer a stirred soil lost 510 grams of water per square foot, while the adjoining compacted soil lost 1680 gram, - a saving due to cultivation of nearly 60 per cent. Wagner, testing the correctness of Nessler's work, found, in 1874, that cultivation reduced the evaporation a little more than 60 per cent; Johnson, 1878, confirmed the truth of the American soils, and Levi Stockbridge, working about the same time, also on American soils, found that cultivation diminished evaporation on a clay soil about 23 per cent, on a sand loam 55 per cent, an on a heavy loam nearly 13 per cent. All the early work done on this subject was done under humid conditions, and it is 0only in recent years that confirmation of this important principle has been obtained for the soils of the dry farm region. Forties, working under California conditions, determined that cultivation reduced the evaporation from the soil surface over 55 per cent. At the Utah Station similar experiments have shown that the saving of soil-moisture by cultivation was 63 per cent for a clay soil, 34 per cent for a coarse sand, and 13 per cent for a clay loam. Further, practical experience has demonstrated time and time again that in cultivation the dry-farmer has a powerful means of preventing evaporation from agricultural soils.

Closely connected with cultivation is the practice of scattering straw or other litter over the ground. Such artificial mulches are very effective in reducing evaporation. Ebermayer found that by spreading straw on the land, the evaporation was reduced 22 per cent; Wagner found under similar conditions a saving of 38 percent, and these results have been confirmed by many other investigators. On the modern dry-farm, which are large in area, the artificial mulching of soils cannot become a very extensive practice, yet it is well to bear the principle in mind. The practice of harvesting dry-farm grain with the header and plowing under the high stubble in the fall is a phase of cultivation for water conservation that deserves special notice. The straw, thus incorporated into the soil, decomposes quite readily in spite of the popular notion to the contrary, and makes the soil more porous, and therefore, more effectively worked for the preservation of evaporation. When this practice is contained for considerable periods, the topsoil becomes rich in organic matter, which assists in retarding evaporation. Besides increasing the fertility of the land. When straw cannot be fed to advantage, as it would be better to scatter it over the land than to burn it, as is often done, anything that over the ground or loosens the topsoil prevent in a measure the evaporation of the water stored in lower soil depths for the use of crops.

Depth of cultivation

The all-important practice for the dry-farmers who is entering upon the growing season is cultivation. The soil must be covered continually with a deep layer of dry loose soil, which because of its looseness and dryness makes evaporation difficult. A leading question in connection with cultivation is the depth to which the soil should be stirred for the best results.

Many of the early students of the subject found that a soil mulch only one half inch in depth was effective in retaining a large part of the soil-moisture which non-cultivated soils would lose by evaporation. Soils differ greatly in the rate of evaporation from their surfaces. Some form a natural mulch when dried, which prevents further water loss. Others form only a thin hard crust below which lies an active evaporating surface of wet soil soils which dry out readily and crumble on top into a natural mulch should be cultivated deeply, for a shallow cultivation does not extend beyond the naturally formed mulch. In fact, on certain calcareous soils, the surfaces of which dry out quickly and form a good protection against evaporation, shallow cultivations often cause a greater evaporation by disturbing the almost perfect natural mulch. Clay or sand soils, which do not so well form a natural mulch, will respond much better to shallow cultivations. In general, however, the deeper the cultivation, the more effective it is in reducing evaporation. Fortier, in the experiments in California to which allusion has already been made, showed the greater value of deep cultivation. During a period of fifteen days, beginning immediately after an irrigation, the soil which had not been mulched lost by evaporation nearly on fourth of the total amount of water that had been added. A mulch 4 inches deep saved about 72 per cent of the evaporation; a mulch 8 inches deep stopped evaporation almost wholly. It is a most serious mistake for the dry farmer, who attempts cultivation for soil moisture conservation, to fail to get the best results simply to save a few cents per acre in added labor.

When to cultivate of till

It has already been shown that the rate of evaporation is greater from a wet than from a dry surface. It follows, therefore, that the critical time for preventing evaporation is when the soil is wettest. After the s soil is tolerably dry, a very large portion of the soil moisture has been lost, which possibly might have been saved by earlier cultivation. The truth of this statement is well shown by experiments conducted by the Utah Station. In one case on a soil well filled with water, during a three week's period, nearly one half of the total loss occurred the first, while only one fifth fell on the third week. Of the amount the first week, over 60 per cent occurred during the first three days. Cultivations should, therefore, be practiced as soon as possible after conditions favorable for evaporation have been established. This means, first, that in early spring, just as soon as the land is dry enough to be worked without causing pudding, the soil should be deeply and thoroughly stirred. Springs plowing, done as early as possible, is an excellent practice for forming a mulch against evaporation. Even when the land has been fall-plowed, spring plowing is very beneficial, though on fall plowed land the disk harrow is usually used in early spring, and if it is set at rather a sharp angle, and properly weighted, so that it cuts deeply into the ground, it is practically as effective as spring plowing. The chief danger to the dry farmer is that he will permit the early spring days to slip by until. When at last he begins spring cultivations, a large portion of the stored soil water has been evaporated. It may be said that deep fall plowing, by permitting the moisture to sing quickly into the lower layer of soil, makes it possible to get upon the ground earlier in the spring. In fact, unplowed land cannot be cultivated as early as that which has gone through the winter in a plowed condition.

If the land carries a fall sown crop, early spring cultivation is doubly important. As soon as the plants are well up in spring the land should be gone over thoroughly several times if necessary, with an iron tooth harrow, the teeth of which are set to slant backward in order not to tear up the plants. The loose earth mulch thus formed is very effective in conserving moisture; and the few plants torn up are more than paid of by the increased water supply for the remaining plants. The wise dry-farmer cultivates his land, whether fallow or cropped as early as possible in the spring.

Following the first spring plowing, disking or cultivation, must come more cultivation. Soon after the spring plowing, the land should be disked and then harrowed. Ever devise should be used to secure the formation of a layer of loose drying soil over the land surface. The season's crop will depend largely upon the effectiveness of this spring treatment.

As the season advances, three causes combine to permit the evaporation of soil-moisture.

First, there is a natural tendency, under the somewhat moist conditions of spring, for the soil to settle compactly and thus to restore the numerous capillary connections with the lower soil layers through which water escapes. Careful watch should therefore be kept upon the soil surface, and whenever the mulch is not loose, the disk or harrow should be run over the land.

Secondly, every rain of spring or summer tends to establish connections with the store of moisture in the soil. In fact, late spring and summer rains are often a disadvantage on dry-farm, which by cultural treatment have been made to containing a large store of moisture. It has been shown repeatedly chat light rains draw moisture very quickly from soil layers many feet below the surface. The rainless summer is not feared by the dry-farmer whose soils are fertile and rich in moisture. It is imperative that at the very earliest moment after a spring or summer rain the topsoil be well stirred to prevent evaporation. It thus happens that in sections of frequent summer rains, as in the Great Plains areas, the farmer has to harrow his land many times in succession, but the increased crop yields invariable justify the added expenditure of effort.

Thirdly, on the summer-fallowed ground weeds start vigorously in the spring and dries upon the soil moisture, if allowed to grow, fully as heavily as a crop of wheat or corn. The dry-farmer must no allow a weed upon his land. Cultivation must be son continuous as to make weeds and impossibility. The belief that the elements added to the soil by weeds offset the loss of soil moisture is wholly erroneous.

The growth of weed on a fallow dry farm in more dangerous than the packed uncared for topsoil. Many implements have been devised of the easy killing of weeds, but none appear to be better than the plow and the disk which are found on very farm. (See Chapter XV)

When crops are growing on the land, through summer cultivation is somewhat more difficult, but must be practiced for the greatest certainty of crop yields. Potatoes, corn, and similar crops may be cultivated with comparative ease, by the use4 of ordinary cultivators. With wheat and the other small grains generally, the damage done to the crop by harrowing late in the season is to great, and reliance is therefore placed on the shading power of the plants to prevent undue evaporation. However, until the wheat and other grains are ten to twelve inches high, it is perfectly safe to harrow them. The teeth should be set backward to diminish the tearing up of the plants, and the implement weighed enough to break the soil crust thoroughly. This practice has been fully tried out over the larger part of the dry-farm territory and found satisfactory.

So vitally important is a permanent soil mulch for the conservation of the water stored in the soil that many attempts have been made to devise means for the effective cultivation of land on which small grains and grasses are growing. In many places plants have been grown in rows so far apart that a man with a hoe could pass between them. Scofield has described this method as practiced successfully in Tunis. Campbell and others in America have proposed that a drill hole be closed every three feet to form a path wide enough for a horse to travel in and to pull a large spring tooth cultivator, with teeth so spaced as to strike between the rows of wheat. It is yet doubtful whether, under average conditions, such careful cultivation, at least of grain crops, is justified by the returns. Under condition of high aridity or where the store of soil moisture is low, such treatment frequently stands between crops success and failure and it is not unlikely that methods will be devised which will permit of the cheap and rapid cultivation between the rows of growing wheat. Meanwhile, the dry-farmer must always remember that the margin under which he works is small, and that his success depends upon the degree to which e prevent s small wasters.

The conservation of soil-moisture depends upon the vigorous, unremitting, continuous stirring of the topsoil. Cultivation! Cultivation! and more cultivation! must be the war-cry of the dry farmer who battles against the water thieves of an arid climate.

CHAPTER IX

REGULATING THE TRANSPIRATION

WATER that has entered the oil may be lost in three ways. First, it may escape by downward seepage, whereby it passes beyond the reach of plant roots and often reaches the standing water. In dry-farm districts such loss is a rare occurrence, for the natural precipitation is not sufficiently large to connect with the country drainage, and it may, therefore, be eliminated from consideration. Second, soil water may be lost by direct evaporation from the surface soil the conditions prevailing in arid districts favor strongly this manner of loss of soil-moisture. It has been shown, however, in the preceding chapter that the farmer by proper and persistent cultivation of the topsoil has it in his power to reduce this loss enough to be almost negligible in the farmers' consideration from the plants themselves. While it is not generally understood, this source of loss is, in districts where dry-farming is properly carried only, very much larger than that resulting either from seepage or from direct evaporation. While plants are growing, evaporation from plants, ordinarily called transpiration, continues. Experiments performed in various arid districts have shown that one and a half to three times more water evaporates from the plant than directly from well-tilled soil. To the most effective methods of checking or controlling this continual loss of water. Transpiration, or the evaporation of water from the plants themselves, and the means of controlling this loss, are subjects of the deepest importance to the dry farmer.

Absorption

To understand the methods for reducing transpiration, as proposed in this chapter, it is necessary to review briefly the manner in which plants take water from the soil. The roots are the organs of water absorption. Practically no water is taken into the pants by the stems or leaves, even under conditions of heavy rainfall. Such small quantities as may enter the plant through the stems and leaves are of very little value in furthering the live and growth of the plant. The roots alone are of real consequence in water absorption the younger roots are most active and effective. Even of the young roots, however, only certain parts are actively engaged in water absorption. At the very tips of the young growing roots are numerous fine hairs, shown largely magnified in Figure 26. These root hairs, which cluster about the growing point of the young roots are the organs of the plant that absorb soil water. They are of value only for limited periods of time for as they grow older, they lose their power of water absorption. In fact they are active only when they are in actual process of growth. It follows, therefore, that water absorption occurs near the tips of the growing roots, and whenever a plant ceases to grow the water absorption ceases also. The root hairs are filled with a dilute substances, as yet poorly understood, which plays an important part in the absorption of water an plant food from the soil.

Fig. 26: Wheat root, showing soil particles clinging to the lower part where the root-hairs are active.

Owing to their minuteness, the root-hairs are in most cases immersed in the water film that surrounds the soil particles, and the soil-water is taken directly into the roots from the soil-water is taken directly into the roots from the soil water film by the process known as osmosis. The explanation of the inward movement is complicated and need into be discussed here. It is sufficient to say that the concentration or strength of the solution with in the root hairs is of different degree from the soil-water solution. The water tends, therefore, to move from the soil into the root, in order to make the solutions inside and outside of the roots of the same concentration. If it should ever occur that the soil-water and the water with in the root hair became the same concentration, that is to say, contained the same substances in the same proportional amounts, their would be no further inward movement of water. Moreover, if it should happen that the soil water is stronger than the water within the root-hair, the water would tend to pass from the plant into the oil this is the condition that prevails in many alkali lands of the West, and is the cause of the death of plants growing o n such lands.

It is clear that under these circumstances not only water enters the root hairs, but many of the substances found in the solution in the soil water enter the plant also. Among these are the mineral substances which are indispensable for the proper life and growth of plants. These plants nutrients are so indispensable that if anyone of them is absent, it is absent it is absolutely impossible for the plant to continue its life functions. The indispensable plant food gathered form the soil by the root hairs, in addition to water are: potassium, calcium, magnesium, iron, nitrogen, and phosphorus, all in their proper combinations. How the plant uses these substances is yet poorly understood, but we are fairly certain that each one has some particular function in the life of the plant. For instance, nitrogen and phosphorus are probably necessary in the formation of the protein or the flesh-forming portion of the plant, while potash is especially valuable in the formation of starch.

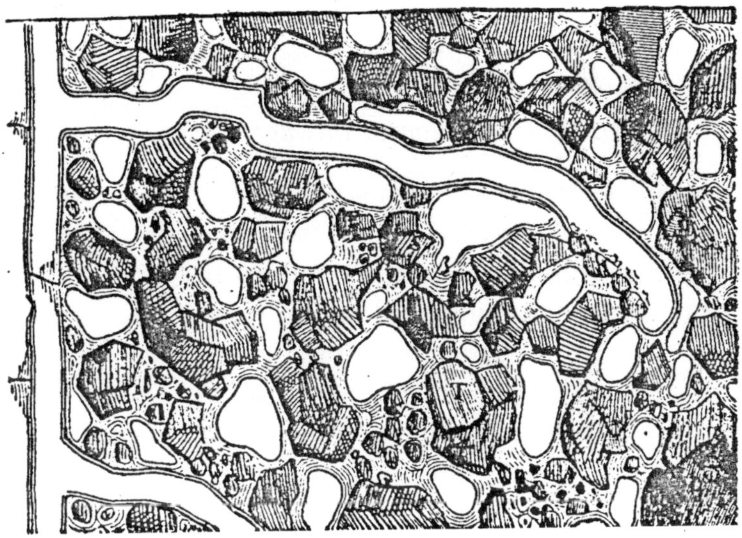

Fig. 27: Penetration of root-hair through soil.

There is a constant movement of the indispensable plant nutrients after they have entered the root-hairs through the stems and into the leaves. This constant movement of the plant foods depends upon the fact that the plant consumers in its own growth considerable quantities of these substances, and as the plant juices are diminished in their content of particular plant foods, more enters form the soil solution. The necessary plant foods do not alone enters the plants, but whatever may be in solution in the soil water enters the plant in the variable quantities. Nevertheless, since the plant uses only a few definite substances and leaves the unnecessary ones in solution, there is soon a cessation of the inward movement of the unimportant constituents of the soil solution. This process is often spoken of as selective absorption; that is the plant, because of it vital activity, appears to have the power of selecting from the soil certain substances rejecting others.

Movement of water through the plant

The soil-water, holding in solution a great variety of plant nutrients, passes from the root-hairs into the adjoining cells and gradually moves from cell to cell throughout the whole plant. in many plants this stream of water does not simply pass from cell to cell, but moves through tubes that apparently have been formed for the specific purpose or aiding the movement of water through the plant. The rapidity of this current is often considerable. Ordinarily, it varies from one foot to six feet per hour, though observations are on record showing that the movement often reached the rate of eighteen feet per hour. It is evident, then, that in an actively growing plant it does not take long for the water which is in the soil to find its way to the uppermost parts of the plant.

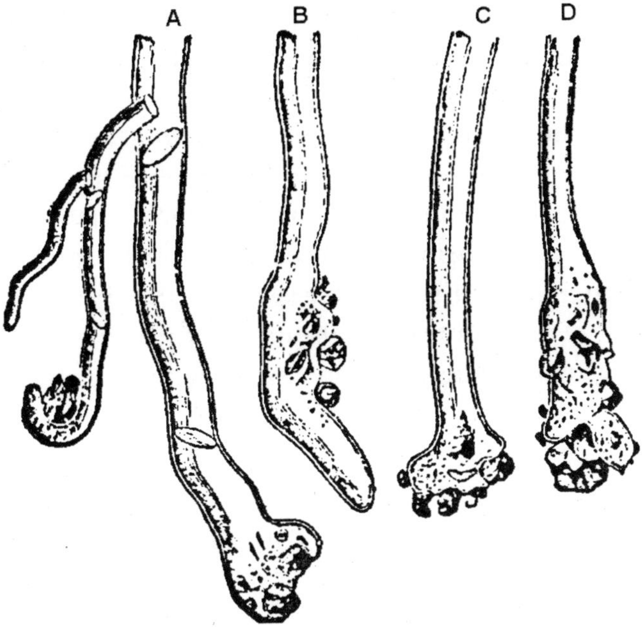

Fig. 28: Magnified root-hairs, showing how soil particles are attached to them.

The work of leaves

Whether water passes upward from cell to cell or through especially provided tubes, it reaches at last the leaves, where evaporation takes place. It is necessary to consider in greater detail what takes place in leaves in order that we may more clearly under stand the loss due to transpiration. One half or more of very plant is made up of the element carbon. The remainder of the plant consists of the mineral substances taken from the soil (not more than two to 10 percent of the dry plant) and water which has been combined with the carbon and these mineral substances to form the characteristic products of plant life. The carbon which forms over half of the plant substances is gathered from the air by the leaves and it is evident that the leaves are very active agents of plant growth. The atmosphere consists chiefly of the gases oxygen and nitrogen in the proportion of one to four, but associated with them are small quantities of various substances. Chief among the secondary constituents of the atmosphere is the gas carbon dioxide, which is formed when carbon burns, that is, when carbon unites with the oxygen of the air. Whenever coal or wood or any carbonaccous substance burns, carbon dioxide is formed, leaves have the power of absorbing the gas carbon dioxide from the air separation the carbon from the oxygen. The oxygen is turned to the atmosphere while the carbon is retained to be used as the fundamental substance in the construction by the plant of oils, fats, starches, sugars, protein and all the other products of plant growth.

This important process known as carbon assimilation is made possible by the aid of countless small openings which exist chiefly on the surfaces of leaves an know as "stomata". The stomata are delicately balanced valves, exceedingly sensitive to external influences. Their appearance under a high power microscope is shown in Figure 29. They are more numerous on the lower side than on the upper side of plants. In fact, there is often five times more on the under side than on the upper side of a leaf. It has been estimated that 150,000 stomata or more are often found per square inch on the under side of the leaves of ordinary cultivated plants.

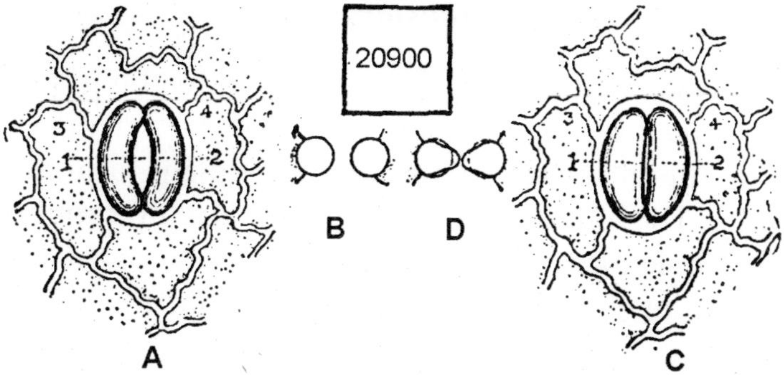

Fig. 29: Diagram of open and partly closed breathing-pores on leaves. Through these openings water escapes from the plant. (From King's "Irrigation and Drainage").

The stomata or breathing pores are so constructed that the may open and close very readily. In wilted leaves they are practically closed; often they also close immediately after a rain: but in strong sunlight they are usually wide open. It is through the stomata that the gases of the air enter the plant through which the discarded oxygen returns to the atmosphere.

It is also through the stomata that the water which is drawn from the soil by the roots through the stems is evaporated into the air. There is some evaporation of water from the stems and the branches of pants, but it is seldom more than a thirtieth or a fortieth of the total transpiration. The evaporation of water from the leaves through the breathing pores is so-called transpiration which is the greatest cause of the soil-water under dry-farm conditions. It is to the prevention of this transpiration that much investigation must be given by future students of dry-farming.

Transpiration

As water evaporates through the breathing pores form the leaves it necessarily follows that a demand is made upon the lower portion of the plant form more water. The effect of the loss of water is felt throughout the whole plant and is, undoubtedly one of the chief causes; of the absorption of water from the soil. As evaporation is diminished the amount of water that enters the plants is also diminished. Yet transpiration appears to be a process wholly necessary for plant life. The question is simply to water extent it may be diminished without injuring plant growth. Many students believe that the carbon assimilation of the plant which is fundamentally important in plant growth, cannot be continued unless there is a steady stream of water passing through the plant and then evaporating form the leaves. Of one thing we are fairly sure: if the upward stream of water is wholly stopped for even a few hours, the plant is likely to be so severely injured as to be greatly handicapped in it future growth.

Botanical authorities agree that transpiration is of value to plant growth, first because it helps to distribute the mineral nutrients necessary for plants growth uniformly through out the plant; secondly, because in permits an active assimilation of the carbon by the leaves; thirdly because it is no unlikely that the heat required to evaporate water, in large part taken from the plant itself, prevent the plant from being overheated. This last mentioned value of transpiration is especially important n dry-farm districts, where, during the summer, the heat is often intense. Fourthly, transpiration in a number of waves not yet clearly understood.

Condition influencing transpiration

In general, the conditions that determine the evaporation of water from the leaves are the same as those that favor the direct evaporation of water from soils, although there seems to be something in the life process of the plant, a physiological factor, which permits or prevents the ordinary water dissipating factors from exercising their full powers. That the evaporation or water from the soil or from a free water surface is not the same as that from plant leaves may be shown in general way from the fact that the amount of water transpired from a give areas of leaf surface may be very much larger or very much smaller that that evaporated from an equal surface of free water exposed to the same conditions. It is further shown by the fact that whereas evaporation from a free water surface goes on with little or no interruption throughout the twenty four hors of the day, transpiration is virtually at stand still at night even though the conditions for the rapid evaporation from a free water surface are present.

Some of the conditions influencing the transpiration may be enumerated as follows:-

First, transpiration influenced by the relative humidity. In dry air under otherwise similar conditions, plants transpire more water than in moist air, though it is to be noted that even when the atmosphere is fully saturated, so that no water evaporates from a free water surface, the transpiration of plants still continues in a small degree. This is explained by the observation that since the life process of a plant produces a certain amount of heat, the plant is always warmer than the surrounding air and that transpiration into an atmosphere fully charged with water vapor is consequently made possible. The fact that transpiration is grater under a low relative humidity is of greatest importance to the dry-farmer, who has to contend with the dry atmosphere.

Second, transpiration increases with increases in temperature; that is, under conditions otherwise the same, transpiration is more rapid on a war day than on a cold one. The temperature increase of itself, however, is not sufficient to cause transpiration.

Third, transpiration increases with the increase of air currents, which is to say, that on a windy day transpiration is much more rapid than on a quiet day.

Fourth, transpiration increases with the increased of direct sunlight. It is an interesting observation that even with the same relative humidity, temperature and wind, transpiration is reduced to a minimum during the night and increases manifold during the day when direct sunlight is available. This condition is again to be noted by the dry farmer, for the dry-farm districts are characterized by and abundance of sunshine.

Fifth, transpiration is decreased by the presence in the soil water of large quantities of the substances which the plant needs for its food material. This will be discussed more fully in the next section.

Sixth, any mechanical vibration of the plant seems to have some effect upon the transpiration. At times it is increased and at times decreased by such mechanical disturbance.

Seventh, transpiration varies also with the age of the plant. In the young plant it is comparatively small. Just before blooming it is very much larger and in time of bloom it is the larges in the history of plant. as the plant grows older transpiration diminishes, and finally at the ripening stage it almost ceases.

Eighth, transpiration varies with the crop. Not all plant take water from the soil at the same rate. Very little is as yet known about the relative water requirements of crops on the basis of transpiration. As an illustration, MacDougall has reported that sagebrush uses about on e fourth as much water as a tomato plant. Even greater differences exist between other plants. This is one of the interesting subjects yet to be investigated by those who are engaged in the reclamation of dry-farm districts. Moreover, the same crop grown under different conditions varies in its rate of transpiration. For instance, plants grown from some time under arid conditions greatly modify their rate of transpiration, as shown by Spalding, who reports that a plant reared under humid condition gave off 3.7 times as much water as the same plant reared under arid conditions. This is very interesting observation tends to confirm the view commonly held that plants grown under arid condition will gradually adapt themselves to the prevailing conditions and is spite of the greater water dissipating conditions will live with the expenditure of less water than would be the case under humid condition. Further, Sorauer found, many years ago, that different varieties of the same crop possess very different rates of transpiration. This also is an interesting subject that should be more fully investigated in the future.

Ninth, the vigor of growth of a crop appears to have a strong influence on transpiration. It does not follow, however that the more vigorously a crop grows the more rapidly does it transpire water, for it is well know that the most luxuriant plant growth occurs in the tropics, where the transpiration is exceedingly low. It seems to be true that under the same conditions, plants that grow most vigorously tend to use the proportionately the smallest amount of water.

Tenth, the root system- its depth and manner of growth- influences the rate of transpiration. They more vigorous and extensive root system, the more rapidly can water be secured from the soil by the plant.

The conditions above enumerated as influencing transpiration are nearly all of a physical character, an it must not be forgotten that they may all be annulled or changed by a physiological regulation. It must be admitted that the subject of transpiration is yet poorly understood, though it is one of the most important subjects in its applications to plant production in localities where water is scarce. It should also be noted that nearly all of the above conditions influencing transpiration are beyond the control of the farmer. The one that seem most readily controlled in ordinary agricultural practice will be discussed in the following section.

Plant food and transpiration

It has been observed repeatedly by students of transpiration that the amount of water which actually evaporates from the leaves is varied materially by the substances held in solution by the soil-eater. That is, transpiration depends upon the nature and concentration of soil solution. This fact, though not commonly applied even at the present time, had really been known for a very long time. Wood ward, in 1699, observed that the amount of water transpired by a plant growing in rain water was 192.3 grams; in spring water, 163.6 grams; that is the amount of water transpired by the plant in the comparatively pure rain water was nearly 20 per cent higher than that used by the plant growing in the notoriously impure water of the River Thames. Sachs, in 1859, carried on an elaborate series of experiments of transpiration in which he showed that the addition of a potassium nitrate, ammonium sulphate or common salt to the solution in which plants grew reduced the transpiration; in fact, the reduction was large, varying from 10 to 75 per cent. This was confirmed by a number of later workers among them for instance, Buergerstein, who, in 1875, showed that whenever acids were added to a soil or to water in which plants are growing, the transpiration is increased greatly; but when alkalies of any kind area added, transpiration decreases. This is of special interest in the development of dry farming, since dry-farms soils, as a rule contain more substances. That m ay be classed as alkalies than do soil maintained under humid conditions. Sour soils are very characteristic of districts where the rainfall is abundant; the vegetation growing on such soils transpires excessively and the crops are consequently more subject to drought.

The investigators of almost a generation ago also determined beyond question that whenever a complete nutrient solution is presented to plants, that is a solution containing all the necessary plant foods. In the proper proportions, the transpiration is reduced immensely. It is not necessary that the plant foods should be presented in a water solution in order to effect this reduction in transpiration; if they are added to the soil on which plants are growing, the same effect will result. The addition of commercial fertilizers to the soil will therefore diminish transpiration it was further discovered nearly half a century ago that similar pants growing on different soil evaporate different amounts of water from their leaves; this difference, undoubtedly, is due to the conditions in the fertility of soils for the more fertile soil is, the richer will the soil-water be in the necessary plant foods. The principle that transpiration or the evaporation of water from the plants depends on the nature and concentration of the soil solution is of far reaching importance in the development of rational practice of dry-farming.

Transpiration for a pound of dry matter

Is plant growth proportion to transpiration? Do plants that evaporate much water grow more rapidly than those that evaporate less? These question s arose very early in the period characterized by an active study of transpiration. If varying the transpiration varies the growth, there would be no special advantage in reducing the transpiration. From an economic point of view the important question is this: Does the plant when its rate of transpiration is reduce still grow with the same vigor? If that be the case, then every effort should be made by the farmer to control and to diminish the rate of transpiration.

One of the very earliest experiments on transpiration, conducted by Woodward in 1699, showed that it requires less water to produce a pound of dry matter if the soil solution were of the proper concentration and contained the elements necessary for plant growth. Little more was done to answer the above question for over one hundred and fifty years. Perhaps the question was not even asked during this period, for scientific agriculture was just coming into being in countries where the rainfall was abundant. However, Tschaplowtiz, in 1878, investigated the subject and found that the increase in dry matter is greatest when the transpiration is the smallest Sorauer in researches conducted form 1880 to 1882, determinate with almost absolute certainty that less water is required to produce a pound of dry matter when the soil is fertilized than when it is not fertilized. Moreover, he observed that the enriching of the soil solution by the addition of artificial fertilizers enabled the plant to produce dry matter with less water. He further found that if a soil is properly tilled so as to set free plant-food an in that way to enrich the soil solution the water –cost of dry plant substances is decreased. Hellriegel, in 1883, confirmed this law and laid down the law that poor plant nutrition increases the water-cost of every pound of dry matter produced. It was about this time that its experiments had shown that during periods of drought the well tilled and well fertilized fields yielded good crops, while the under fertilized fields yielded poor crops or crop failure- indicating thereby, since rainfall was the critical factor, that the fertility of the soil is important in determining whether or not with a small amount of water a good crop can be produced. Pagnoul working in 1895 with fescue grass, arrived at the same conclusion. On poor clay soil it required 1109 pound of water to produced one pound of dry matter, while on a rich calcareous soil only 574 pound were required. Gardner of the United States Department of Agriculture, Bureau of Soils, working in 1908, on the manuring of soils came to the conclusion that the more fertile the soil the less water is required to produce a pound of dry matter, he incidentally called attention to the fact that in countries of limited rainfall this might be a very important principle to apply in crop production. Hopkins in his study of the soils of Illinois has repeatedly observed, in connection with certain soils, that where the land is kept fertile, injury from drought is not common, implying thereby that fertile soils will produce dry matter at a lower water cost. The most recent experiments on this subject conducted by Utah station, confirm theses conclusions. The experiments, which covered several years, were conducted in pots filled with different soil on a soil, naturally fertile, 908 pounds of water were transpired for each pound of dry matter (Corn) produced; by adding to this soil an ordinary dressing of manure, this was reduced to 613 pounds and by adding small amount of sodium nitrate it was reduced to 585 pounds. If so large a reduction could be secured in practice, it would seem to justify the use of commercial fertilizers in years when the dry-farm year opens with little water stored in the soil. Similar results, as will be shown below, were obtained by the use of various cultural methods. In may, therefore, be stated as a law, that any cultural treatment which enables the soil-water to acquire larger quantities of plant-food also enables the plant to produce dry matter with the use of a smaller amount of water. In dry-farming where the limiting factor is water, this principle must be emphasized in every cultural operation.

Methods of controlling transpiration

It would appear that at present the only means possessed by the farmer for controlling transpiration and making possible maximum crops with the minimum amount of water in a properly tilled soil is to keep the soil as fertile as is possible. In the light of this principle the practices already recommended for the storing of water and for the prevention of the direct evaporation of water from the soil are again emphazied. Deep and frequent plowing, preferably in the fall so that the weathering of the winter may be felt deeply and strongly, is of first importance in liberating plant-food. Cultivation which has been recommended for the prevention of the direct evaporation of water is of itself an effective factoring setting free plant-food and thus in reducing the amount of water required by plants. The experiments at the Utah Station, already referred to, bring out very strikingly the value of cultivation in reducing the transpiration. For instance, in a series of experiments the following results were obtained. On a sandy loam, not cultivated, 603 pounds of water were transpired to produce one pound of dry matter of corn; on the same soil, cultivated, only 252 pounds were required. On a clay loam, not cultivated, 535 pounds of water were transpired for each pound of dry matter, whereas on the cultivated soil only 428 pounds were necessary. On a clay soil, not cultivated, 753 pounds of water were transpired for each pound of dry matter; on the cultivated soil, only 582 pounds. The farmer who faithfully cultivates the soil throughout the summer and after every rain has therefore the satisfaction of knowing that he is accomplishing two very important things: he is keeping the moisture in the soil, and he is making it possible for good crops to be grown with much less after than would otherwise be required. Even in the case of a peculiar soil on which ordinary cultivation did not reduce the direct evaporation, the effect upon the transpiration was very marked. On the soil which was not cultivated 451 pounds of water were required to produce one pound of dry matter (corn), while on the cultivated soils, though the direct evaporation was no smaller, the number of pounds of water for each pound of dry substance was as low as 265.

One of the chief values of fallowing lies in the liberation of the plant-food during the fallow year, which reduces the quantity of water required the next year for the full growth of crops. The Utah experiments to which reference has already been made show the effect of the previous soil treatment upon the water requirements of crops. One half of the three types of soil had been cropped for three successive years, while the other half had been left bare. During the fourth year both halves were planted to corn. For the sandy loam it was found that, on the part that had been cropped previously, 659 pounds of water were required for each pound of dry matter produced, while on the part that had been bare only 573 pounds were required. For the clay loam 889 pounds on the cropped part and 550 on the previously bare part were required for each pound of dry matter. For the clay 7466 pounds on the cropped part and 1739 pounds on the previously bare part were required for each pound of dry matter. These results teach clearly and emphatically that the fertile condition of the soil induced by fallowing makes it possible to produce dry matter with a smaller amount of water than can be done on soils that are cropped continuously. The beneficial effects of fallowing are therefore clearly twofold: to store the moisture of two seasons for the use of one crop; and to set free fertility to enable the plant to grow with the least amount of water. It is not yet fully understood what changes occur in fallowing to give the soil the fertility which reduces the water needs of the plant. The researches of Atkinson in Mnotana, Stewart and Graves in Utah, and Jensen in South Dakota make it seem probable that the formation of nitrates plays an important part in the whole process.

If a soil is of such a nature that neither careful deep plowing at the right time nor constant crust cultivation are sufficient to set free an abundance of plant-food, it maybe necessary to apply manures or commercial fertilizers to the soil. While the question of restoring soil fertility has not yet come to be a leading one in dry-farming, yet in view of what has been said in this chapter it is not impossible that the time will come when the farmers must give primary attention to soil fertility in addition to the storing and conservation of soil-moisture. The fertilizing of lands with proper plant-foods, as shown in the last sections, tends to check transpiration and makes possible the production of dry matter at the lowest water-cost.

The recent practice in practically all dry-farm districts, at least in the intermountain and far West, to use the header for harvesting bears directly upon the subject considered in this chapter. The high stubble which remains contains much valuable plant-food, often gathered many feet below the surface by the plant roots. When this stubble is plowed under there is a valuable addition of the plant-food to the upper soil. Further, as the stubble decays, acid substances are produced that act upon the son grains to set free the plant –food locked up in them. The plowing under of stubble is therefore of great value to the dry-farmer. The plowing under of any other organic substance has the same effect. In both cases fertility is concentrated near the surface, which dissolves in the soil-water and enables the crop to mature with the least quantity of water.

The lesson then to be learned from this chapter is, that it is not sufficient for the dry-farmer to store an abundance of water in the soil and to prevent that water from evaporating directly from the soil; but the soil must be kept in such a state of high fertility that plants are enabled to utilize the stored moisture in the most economical manner. Water storage, the prevention of evaporation, and the maintenance of soil fertility go hand in hand in the development of a successful system of farming without irrigation.

CHAPTER X

PLOWING AND FALLOWING

The soil treatment prescribed in the preceding chapters rests upon

(1) deep and thorough plowing, done preferably in the fall;

(2) thorough cultivation to form a mulch over the surface of the land, and

(3) clean summer fallowing every other year under low rainfall or every third or fourth year under abundant rainfall.

Students of dry-farming all agree that thorough cultivation of the topsoil prevents the evaporation of soil-moisture, but some have questioned the value of deep and fall plowing and the occasional clean summer fallow. It is the purpose of this chapter to state the findings of practical men with reference to the value of plowing and fallowing in producing large crop yields under dry-farm conditions.

It will be shown in Chapter XVIII that the first attempts to produce crops without irrigation under a limited rainfall were made independently in many diverse places. California, Utah and the Columbia Basin, as far as can now be learned, as well as the Great Plains area, were all independent pioneers in the art of dry-farming. It is a most significant fact that these diverse localities, operating under different conditions as to soil and climate, have developed practically the same system of dry-farming. In all these places the best dry-farmers practice deep plowing wherever the subsoil will permit it; fall plowing wherever the climate will permit it; the sowing of fall grain wherever the winters will permit it, and the clean summer fallow every other year, or every third or fourth year. H.W. Campbell, who has been the leading exponent of dry-farming in the great Plains area, began his work without the clean summer fallow as a part of his system, but has long since adopted it for that section of the country. It is scarcely to be believed that these practices, developed laboriously through a long succession of years in widely separated localities, do not rest upon correct scientific principles. In any case, the accumulated experience of the dry-farmers in this country confirms the doctrines of soil tillage for dry-farms laid down in the preceding chapters.

At the Dry-Farming Congresses large numbers of practical farmers assemble for the purpose of exchanging experiences and views. The reports of the Congress show a great difference of opinion on minor matters and a wonderful unanimity of opinion on the more fundamental questions. For instance, deep plowing was recommended by all who touched upon the subject in their remarks; though one farmer, who lives in a locality the subsoil of which was very inert, recommended that the depth of plowing should be increased gradually until the full depth is reached, to avoid a succession of poor crop years while the lifeless soil was being vivified. The states of Utah, Montana, Wyoming, South Dakota, Colorado, Kansas, Nebraska, and the provinces of Alberta and Saskatchewan Canada all specifically declared through one to eight representatives from each state in favor of deep plowing as a fundamental practice in dry-farming. Fall plowing, wherever the climatic conditions make it possible, was similarly advocated by all the speakers. Farmers in certain localities had found the soil so dry in the fall that plowing was difficult, but Campbell insisted that even in such places it would be profitable to use power enough to break up the land before the winter season set in. Numerous speakers from the stats of Utah, Wyoming, Montana, Nebraska, and a number of the Great Plains states, as well as from the Chinese Empire, declared themselves as favoring fall plowing. Scarcely a dissenting voice was raised.

In the discussion of the clean summer fallow as a vital principle of dry-farming a slight difference of opinion was discovered. Farmers from some of the localities insisted that the clean summer fallow every other year was indispensable; others that one in three years was sufficient; and others one in four years and a few doubted the wisdom of its altogether. However, all the speakers agreed that clean and thorough cultivation should be practiced faithfully during the spring, summer and fall of the fallow year. The appreciation of the fact that weds consume precious moisture and fertility seemed to be general among the dry-farmers from all sections of the country. The following states, provinces, and countries declared themselves as being definitely and emphatically in favor of clean summer fallowing: California, Utah, Nevada, Washington, Montana, Idaho, Colorado, New Mexico, North Dakota, Nebraska, Alberta, Saskatchewan, Russia, Turkey, the Transvaal, Brazil and Australia. Each of these many districts was represented by one to ten or more representatives. The only state to declare somewhat vigorously against it was from the Great plains area, and a warning voice was heard from the United States Department of Agriculture. The recorded practical experience of the farmers over the whole of the dry-farm territory of the United States leads to the conviction that fallowing must be accepted as a practice which resulted in successful dry-farming. Further, the experimental leaders in the dry-farm movement, whether working under private, state, or governmental direction, are, with very few exceptions, strongly in favor of deep fall plowing and clean summer fallowing as parts of the dry-farm system.

The chief reluctance to accept clean summer fallowing as a principle of dry-farming appears chiefly among students of the Great Plains area. Even there it is admitted by all that a wheat crop following a fallow year is larger and better than one following wheat. There seem, however, to be two serious reasons for objecting to it. First, a fear that a clean summer fallow, practiced every second, third, or fourth year, will cause a large diminution of the organic matter in the soil, resulting finally in complete crop failure; and second, a belief that a hoed crop, like corn or potatoes, exerts the same beneficial effect.

It is undoubtedly true that the thorough tillage involved in dry-farming exposes to the action of the elements the organic matter of the soil and thereby favors rapid oxidation. For that reason the different ways in which organic matter may be supplied regularly to dry-farms are pointed out in Chapter XIV. It may also be observed that the header harvesting system employed over a large part of the dry-farm territory leaves the large header stubble to be plowed under, and it is probable that under such methods more organic matter is added to the soil during the year of cropping than is lost during the year of fallowing. It may, moreover, be observed that thorough tillage of a crop like corn or potatoes tends to cause a loss of the organic matter of the soil to a degree nearly as large as is the case when a fallow field is well cultivated. The thorough stirring of the soil under an arid or semiarid climate, which is an essential feature of dry-farming, will always result in a decrease in organic matter. It matters little whether the soil is fallow or in crop during the process of cultivation, so far as the result is concerned.

A serious matter connected with fallowing in the Great Plains area is the blowing of the loose well-tilled soil of the fallow fields, which results from the heavy winds that blow so steadily over a large part of the western slope of the Mississippi Valley. This is largely avoided when crops are grown on the land, even when it is well tilled.

The theory, recently proposed, that in the Great Plains area, where the rains come chiefly in summer, the growing of hoed crops may take the place of the summer fallow, is said to be based on experimental data not yet published. Careful and conscientious experimenters, as Chilcott and his co-laborers, indicate in their statements that in many cases the yields of wheat after a hoed crop, have been larger than after a fallow year. The doctrine has, therefore, been rather widely disseminated that fallowing has no place in the dry-farming of the Great Plains area and should be replaced by the growing of hoed crops. Chilcott, who is the chief exponent of this doctrine, declares, however, that it is only with spring-grown crops and for a succession of normal years that fallowing may be omitted, and that fallowing must be resorted to as a safeguard or temporary expedient to guard against total loss of crop where extreme drought is anticipated; that is, where the rainfall falls below the average. He further explains that continuous grain cropping, even with careful plowing and spring and fall tillage, is unsuccessful; but holds that certain rotations of crops, including grain and a hoed crop every other year, are often more profitable than grain alternating with clean summer fallow. He further believes that the fallow year every third or fourth year is sufficient for Great Plains conditions.

Jardine explains that whenever fall grain is grown in the Great Plains area, the fallow is remarkably helpful and in fact because of the dry winters is practically indispensable.

This latter view is confirmed by the experimental results obtained by Atkinson and others at the Montana Experiment Stations, which are conducted under approximately Great Plains conditions. The average results follow:-

TABLE: 13

Substation	Kubanka spring wheat		White hull less barley		Sixty-day oats	
	Grown continuously	After fallow	Grown continuously	After fallow	Grown continuously	After fallow
	Bushel	Bushel	Bushel	Bushel	Bushel	Bushel
Dawson County	15.18	17.57	15.97	20.90	31.17	51.00
Rosebud County	16.98	20.80	15.02	28.31	30.21	40.03
Yellowstone County	7.73	19.32	14.90	20.33	13.75	47.94
Chouteau County	14.18	17.35	13.29	11.95	28.90	34.56
Average	13.52	18.76	14.79	20.37	26.01	43.38

It should be mentioned also that in Saskatchewan, in the north end of the Great Plains area, and which is characteristic, except for a lower annual temperature, of the whole area, and where dry-farming has been practiced for a quarter of a century, the clean summer fallow has come to be an established practice.

This recent discussion of the place of fallowing in the agriculture of the Great Plains area illustrates what has been said so often in this volume about the adapting of principles to local conditions. Wherever the summer rainfall is sufficient to mature a crop, fallowing for the purpose of storing moisture in the soil is unnecessary; the only value of the fallow year under such conditions would be to set free fertility. In the Great Plains area the rainfall is somewhat higher than elsewhere in the dry-farm territory and most of it comes in summer; and the summer precipitation is probably enough in average years to mature crops, providing soil conditions are favorable. The main considerations, then, are to keep the soils open for the reception of the water and to maintain the soils in a sufficiently fertile condition to produce, as explained in Chapter IX, plants with a minimum amount of water. This is accomplished very large by the year of hoed crop, when the soil is as well stirred as under a clean fallow.

The dry-farmer must never forget that the critical element in dry-farming is water and that the annual rainfall will in the very nature of things vary from year to year, with the result that the dry year, or the year with a precipitation below the average, is sure to come. In somewhat wet years the moisture stored in the soil is of comparatively little consequence, but in year of drought it will be the main dependence of the farmer. Now, whether a crop be hoed or not, it requires water for its growth, and land which is continuously cropped even with a variety of crops is likely to be so largely depleted of its moisture that, when the year of drought comes, failure will probably result.

The precariousness of dry-farming must be done away with. The year of drought must be expected every year. Only as certainty of crop yield is assured will dry-farming rise to a respected place by the side of other branches of agriculture. The attain such certainty and respect clean summer fallowing every second, third, or fourth year, according to the average rainfall, is probably indispensable; and future investigations, long enough continued, will doubtless confirm this prediction. Undoubtedly, a rotation of crops, including hoed crops, will find an important place in dry-farming, but probably not to the complete exclusion of the clean summer fallow.

Jethro Tull, two hundred years ago, discovered that thorough tillage of the soil gave crops that in some cases could not be produced by the addition of manure, and he came to the erroneous conclusion that "tillage is manure". In recent days we have learned the value of tillage in conserving moisture and in enabling plants to reach maturity with the least amount of water, and we may be tempted to believe that "tillage is moisture". This, like Tull's statement, is a fallacy and must be avoided. Tillage can take the place of moisture only to a limited degree. Water is the essential consideration in dry-farming, else there would be no dry-farming.

CHAPTER XI

SOWING AND HARVESTING

The careful application of the principles of soil treatment discussed in the preceding chapters will leave the soil in good condition for sowing, either in the fall or spring. Nevertheless, though proper dry-farming insures a first-class seed-bed, the problem of sowing is one of the most difficult in the successful production of crops without irrigation. This is chiefly due to the difficulty of choosing, under somewhat rainless conditions, a time for swing that will insure rapid and complete germination and the establishment of a root system capable of producing good plants. In some respects fewer definite, reliable principles can be laid down concerning sowing than any other principle of important application in the practice of dry-farming. The experience of the last fifteen years has taught that the occasional failures to which even good dry-farmers have been subjected have been caused almost wholly by uncontrollable unfavorable conditions prevailing at the time of sowing.

Conditions of germination

Three conditions determine germination: (1) heat, (2) oxygen, and (3) water. Unless these three conditions are all favorable, seeds cannot germinate properly. The first requisite for successful seed germination is a proper degree of heat. For every kind of seed there is a temperature below which germination does not occur; another, above which it does not occur, and another, the best at which, providing the other factors are favorable, germination will go on most rapidly. The following table, constructed by Goodale, shows the latest, highest and best germination temperatures for wheat, barley, and corn. Other seeds germinate approximately within the same ranges of temperature:-

TABLE: 14
GERMINATION TEMPERATURES
(DEGREES FAHRENHEIT).

	Lowest	Highest	Best
Wheat	41	108	84
Barley	41	100	84
Corn	49	115	91

Germination occurs within the considerable range between the highest and lowest temperatures of this table, though the rapidity of germination decreases as the temperature recedes from the best. This explains the early spring and late fall germination when the temperature is comparatively low. If the temperature falls below the lowest required for germination, dry seeds are not injured, and even a temperature far below the freezing point of water will not affect seeds unfavorably if they are not too moist. The warmth of the soil, essential to germination, cannot well be controlled by the farmer; and planting must, therefore, be done in season when, from past experience, it is probable that the temperature is and will remain in the neighborhood of the best degree for germination. More heat is required to raise the temperature of wet soils; therefore, seeds will generally germinate more slowly in wet than in dry soils, as is illustrated in the rapid germination often observed in well-tilled dry-farm soils. Consequently, it is safer at a low temperature to sow in dry soils than in wet ones. Dark soils absorb heat more rapidly than lighter colored ones, and under the same conditions of temperature germination is therefore more likely to go on rapidly in dark colored soils. Over the dry-farm territory the soils are generally light colored, which would tend to delay germination. The incorporation of organic matter with the soil, which tends to darken the soil, has a slight though important bearing on germination as well as on the general fertility of the soil, and should be made an important dry farm practice. Meanwhile, the temperature of the soil depends almost wholly upon the prevailing temperature conditions in the district and is not to any material degree under the control of the farmer.

A sufficient supply of oxygen in the soil is indispensable to germination. Oxygen, as is well known, forms about one fifth of the atmosphere and is the active principle in combustion and in the changes in the animal body occasioned by respiration. Oxygen should be present in the soil air in approximately the proportion in which it is found in the atmosphere. Germination is hindered by a larger or smaller proportion than is found in the atmosphere. The soil must be in such a condition that the air can easily enter or leave the upper soil layer; that is, the soil must be somewhat loose. In order that the seeds may have access to the necessary oxygen, then, sowing should not be done in wet or packed soils, nor should the sowing implements be such as to press the soil too closely around the seeds. Well-fallowed soil is in an ideal condition for admitting oxygen.

If the temperature is right, germination begins by the forcible absorption of water by the seed from the surrounding soil. The force of this absorption is very great, ranging from four hundred to five hundred pounds pre square inch, and continues until the seed is completely saturated. The great vigor with which water is thus absorbed from the soil explains how seeds are able to secure the necessary water from the thin water film surrounding the soil grains. The following table, based upon numerous investigations conducted in Germany and Utah, shows the maximum percentages of water contained by seeds when the absorption is complete. These quantities are reached only when water is easily accessible:-

TABLE: 15
PERCENTAGE OF WATER CONTAINED BY SEEDS AT SATURATION

	German	Utah
Rye	58	-
Wheat	57	52
Oats	58	43
Barley	50	44
Corn	44	57
Peas	93	-
Beans	95	88
Lucern	78	87

Germination itself does not go on freely until this maximum saturation has been reached. Therefore, if the moisture in the soil is low, the absorption of water is made difficult and germination is retarded. This shows itself in a decreased percentage of germination. The effect upon germination of the percentage of water in the soil is well shown by some of the Utah experiments, as follows:-

TABLE: 16
EFFECT OF VARYING AMOUNTS OF WATER PERCENTAGE OF GERMINATION

Per cent water in soil	7.5	10	12.5	15	17.5	30	22.5	25
Wheat in Sandy Loam	0.0	98	94.0	86	82.0	82	82.0	6
Wheat in Clay	30.0	48	84.0	94	84.0	82	86.0	58
Bean sin Sandy Loam	0.0	0	20.0	46	66.0	18	8.0	9
Beans in Clay	0.0	0	6.0	20	22.0	32	30.0	36
Lucern in Sandy Loam	0.0	18	68.0	54	54.0	8	8.0	9
Lucern in Clay	8.0	8	54.0	48	50.0	32	14.9	14

In a sandy soil a small percentage of water will cause better germination than in a clay soil. While different seeds vary in their power to abstract water from soils, yet it seems that for the majority of plants, the best percentage of soil-water of germination purposes is that which is in the neighborhood of the maximum field capacity of soils of water, as explained in Chapter VII. Bogdanoff has estimated that the best amount of water in the soil for germination purposes is about twice the maximum percentage of hygroscopic water. This would not be far from the field-water capacity as described in the preceding chapter.

During the absorption of water, seeds swell considerably, in many cases from two to three times their normal size. This has the very desirable effect of crowding the seed walls against the soil particles and thus, by establishing more points of contact, enabling the seed to absorb moisture with greater facility. As seeds begin to absorb water, heat is also produced. In many cases the temperature surrounding the seeds is increased one degree on the Centigrade scale by the mere process of water absorption. This favors rapid germination. Moreover, the fertility of the soil has a direct influence upon germination. In fertile soils the germination is more rapid and more complete than in infertile soils. Especially active in favoring direct germination are the nitrates. When it is recalled that the constant cultivation and well-kept summer fallow of dry-farming develop large quantities of nitrates in the soil, it will be understood that the methods of dry-farming as already outlined accelerate germination very greatly.

It scarcely need be said that the soil of the seedbed should be fine, mellow, and uniform in physical texture so that the seeds can be planted evenly and in close contact with the soil particles. All the requisite conditions for germination are best met by the conditions prevailing in a well-kept summer fallowed soil.

Time to sow

In the consideration of the time to sow, the first question to be disposed of by the dry-farmer is that of fall as against spring sowing. The small grains occur as fall and spring varieties, and it is vitally important to determine which season, under dry-farm conditions, is the best for sowing.

The advantages of fall sowing are many. As stated, successful germination is favored by the presence of an abundance of fertility, especially of nitrates, in the soil. In summer-fallowed land nitrates are always found in abundance in the fall, ready to stimulate the seed into rapid germination and the young plants into vigorous growth. During the late fall and winter months the nitrates disappear, at least in part, and from the point of view of fertility the spring is not so desirable as the fall for germination. More important, grain sown in the fall under favorable conditions will establish a good root system which is ready for use and in action in the early spring as soon as the temperature is right and long before the farmer can go out on the ground with his implements. As a result, the crop has the use of the early spring moisture, which under the conditions of spring sowing is evaporated into the air. Where the natural precipitation is light and the amount of water stored in the soil is not large, the gain resulting from the use of the early spring moisture often decides the question in favor of fall sowing.

The disadvantages of fall sowing are also many. The uncertainty of the fall rains must first be considered. In ordinary practice, seed sown in the fall does not germinate until a rain comes, unless indeed sowing is done immediately after a rain. The fall rains are uncertain as to quantity. In many cases they are so light that they suffice only to start germination and not to complete it and give the plants the proper start. Such incomplete germination frequently causes the total loss of the crop. Even if the stand of the fall crop is satisfactory, there is always the danger of winter-killing to be reckoned with. The real cause of winter-killing is not yet clearly understood, though it seems that repeated thawing and freezing, drying winter winds, accompanied by dry cold or protracted period of intense cold, destroy the vitality of the seed and young root system. Continuous but moderate cold is not ordinarily very injurious. The liability to winter-killing is, therefore, very much greater wherever the winters are open than in places where the snow covers the ground the larger part of the winter. It is also to be kept in mind that some varieties are very resistant to winter-killing, while others require well-covered winters. Fall sowing is preferable wherever the bulk of the precipitation comes in winter and spring and where the winters are covered for some time with snow and the summers are dry. Under such conditions it is very important that the crop make use of the moisture stored in the soil in the early spring. Wherever the precipitation comes largely in late spring and summer, the arguments in favor of fall sowing are not so strong, and in such localities spring sowing is often more desirable than fall sowing. In the Great Plains district, therefore, spring sowing is usually recommended, though fall-sown crops nearly always, even there, yield the larger corps. In the intermountain states, with wet winters and dry summers, fall sowing has almost wholly replaced spring sowing. In fact, Farrell reports that upon the Nephi (Utah) substation the average of six years shows about twenty bushels of wheat from fall-sown seed as against about thirteen bushels from spring-sown seed. Under the California climate, with wet winters and a winter temperature high enough for plant growth, fall sowing is also a general practice. Wherever the conditions are favorable, fall sowing should be practiced, for it is in harmony with the best principles of water conservation. Even in districts where the precipitation comes chiefly in the summer, it may be found that fall sowing, after all, is preferable.

The right time to sow in the fall can be fixed only with great difficulty, for so much depends upon the climatic conditions. In fact the practice varies in accordance with differences in fall precipitation and early fall frosts. Where numerous fall rains maintain the soil in a fairly moist condition and the temperature is not too low, the problem is comparatively simple. In such districts, for latitudes represented by the dry-farm sections of the United States, a good time for all planting is ordinarily from the first of September to the middle of October. If sown much earlier in such districts, the growth is likely to be too rank and subject to dangerous injury by frosts, and as suggested by Farrell the very large development of the root system in the fall may cause, the following summer, dangerously large growth of foliage; that is, the crop may run to straw at the expense of the grain. If sown much later, the chances are that the crop will not possess sufficient vitality to withstand the cold of late fall and winter. In localities where the late summer and the early fall are rainless, it is much more difficult to lay down a definite rule covering the time of fall sowing. The dry-farmers in such places usually sow at any convenient time in the hope that an early rain will start the process of germination and growth. In other cases planting is delayed until the arrival of the first fall rain. This is an uncertain and usually unsatisfactory practice, since if often happens that the sowing is delayed until to late in the fall for the best results.

In districts of dry late summer and fall, the greatest danger in depending upon the fall rains for germination lies in the fact that the precipitation is often so small that is initiates germination without being sufficient to complete it. This means that when the seed is well started in germination, the moisture gives out. When another slight rain comes a little later, germination is against started and possibly again stopped. In some seasons this may occur several times, to the permanent injury of the crop. Dry-farmers try to provide against this danger by using an unusually large amount of seed, assuming that a certain amount will fail to come up because of the repeated partial germinations. A number of investigators have demonstrated that a seed may start to germinate, then be dried, and again be started to germinate several times in succession without wholly destroying the vitality of the seed. Nowoezek has conducted a number of experiments on this subject, with the following results:-

TABLE: 17
EFFECT OF REPEATED DRYING ON PERCENTAGE OF GERMINATION

	First germination	Third germination	Fifth germination	Seventh germination
Wheat	75	57	25	1
Barley	85	74	33	4
Oats	90	77	40	8
Corn	98	66	3	0
Peas	87	3	0	0

In these experiments wheat and other seeds were allowed to germinate and dry seven times in succession. With each partial germination the percentage of total germination decreased until at the seventh germination only a few seeds of wheat, barley, and oats retained their power. This, however, is practically the condition in dry-farm districts with rainless summers and falls, where fall seeding is practiced. In such localities little dependence should be placed on the fall rains and greater reliance placed on a method of soil treatment that will insure good germination. For this purpose the summer fallow has been demonstrated to be the most desirable practice. If the soil has been treated according to the principles laid down in earlier chapters, the fallowed land will, in the fall, contain a sufficient amount of moisture to produce complete germination though no rains may fall. Under such conditions the main consideration is to plant the seed so deep that it may draw freely upon the stored soil-moisture. This method makes fall germination sure in districts where the natural precipitation is not to be depended upon.

When sowing is done in the spring, there are few factors to consider. Whenever the temperature is right and the soil has dried out sufficiently so that agricultural implements may be used properly, it is usually safe to begin sowing. The customs which prevail generally with regard to the time of spring sowing may be adopted in dry-farm practices also.

Depth of seeding

The depth to which seed should be planted in the soil is of importance in a system of dry-farming. The reserve materials in seeds are used to produce the first roots and the young plants. No new nutriment beyond that stored in the soil can be obtained by the plant until the leaves are above the ground, able to gather carbon from the atmosphere. The danger of deep planting lies, therefore, in exhausting the reserve materials of the seeds before the plant has been able to push its leaves above the ground. Should this occur, the plant will probably die in the soil. On the other hand, if the seed is not planted deeply enough, it may happen that the roots cannot be sent down far enough to connect with the soil-water reservoir below. Then, the root system will not be strong and deep, but will have to depend for its development upon the surface water, which is always a dangerous practice in dry-farming. The rule as to the depth of seeding is simply: Plant as deeply as is safe. The depth to which seeds may be safely placed depends upon the nature of the soil, its fertility, its physical condition, and the water that it contains. In sandy soils, planting maybe deeper than in clay soils, for it requires less energy for a plant to push roots, stems, and leaves through the loose sandy soil than through the more compact clay soil; in a dry soil planting maybe deeper than in wet soils; likewise, deep planting is safer in a loose soil than in one firmly compacted; finally, where the moist soil is near the surface. Countless experiments have been conducted on the subject of depth of seeding. In a few cases, ordinary agricultural seeds planted eight inches deep have come up and produced satisfactory plants. However, the consensus of opinion is that from one to three inches are besting humid districts, but that everything considered, four inches is the best depth under dry-farm conditions. Under a low natural precipitation, where the methods of dry-farming are practiced, it is always safe to plant deeply, for such a practice will develop and strengthen the root system, which is one big step toward successful dry-farming.

Quantity to sow

Numerous dry-farm failures may be charged wholly to ignorance concerning the quantity of seed to sow. In no other practice has the custom of humid countries been followed more religiously by dry-farmers, and failure has nearly always resulted. The discussion in this volume have brought out the fact that every plant of whatever character requires a large amount of water for its growth. From the first day of its growth to the day of its maturity, large amount of water are taken from the soil through the plant and evaporated into the air through the leaves. When the large quantities of seed employed in humid countries have been sown on dry lands, the result has usually been an excellent stand early in the season, with a crop splendid in appearance up to early summer. A luxuriant spring crop reduces, however, the water content of the soil so greatly than when the heat of the summer arrives, there is not sufficient water left in the soil to support the final development and ripening. A thick stand in early spring is no assurance to the dry-farmer of a good harvest. On the contrary, it is usually the field with a thin stand in spring that stands up best through the summer and yields most at the time of harvest. The quantity of seed sown should vary with the soil conditions: the more fertile the soil is, the more seed may be used; the more water in the soil, the more seed may be sown; as the fertility or the water content diminishes, the amount of seed should likewise be diminished. Under dry-farm conditions the fertility is good, but the moisture is low. As a general principle, therefore, light seeding should be practiced on dry-farms, though it should be sufficient to yield a crop that will shade the ground well. If the sowing is done early, in fall or spring, less seed maybe used than if the sowing is late, because the early sowing gives a better chance for root development, which results, ordinarily, in more vigorous plants that consume more moisture than the smaller and weaker plants of later sowing. If the winters are mild and well covered with snow, less seed may be used than in districts where severe or open winters cause a certain amount of winter-killing. On a good seed-bed of fallowed soil less seed may be used than where the soil has not been carefully tilled and is somewhat rough and lumpy and unfavorable for complete germination. The yield of any crop is not directly proportional to the amount of sown, unless all factors contributing to germination are alike. In the case of wheat and other grains, thin seeding also gives a plant a better chance for stooling, which Nature's method of adapting the plant to the prevailing moisture and fertility conditions. When plants are crowded, stooling cannot occur to any marked degree, and the crop is rendered helpless in attempts to adapt itself to surrounding conditions.

In general the rule maybe laid down that a little more than one half as much seed should be used in dry-farm districts with an annual rainfall of about fifteen inches than is used in humid districts. That is as against the customary five pecks of wheat used per acre in humid countries about three pecks or even two pecks should be used on dry-farms. Merrill recommends the seeding of oats at the rate of about three pecks per acre; of barley, about three pecks; of rye, two pecks; of alfalfa, six pounds; of corn, two kernels to the hill, and other crops in the same proportion. No invariable rule can be laid down for perfect germination. A small quantity of seed is usually sufficient; but where germination frequently fails in part, more seed must be used. If the stand is too thick at the beginning of the growing season, it must be harrowed out. Naturally, the quantity of seed to be used should be based on the number of kernels as well as on the weight. For instance, since the larger the individual wheat kernels the fewer in a bushel, fewer plants would be produced from a bushel of large than from a bushel of small seed wheat. The size of the seed in determining the amount for sowing is often important and should be determined by some simple method, such as counting the seeds required to fill a small bottle.

Method of sowing

There should really be no need of discussing the method of sowing were it not that even at this day there are farmers in the dry-farm district who sow by broadcasting and insist upon the superiority of this method. The broadcasting of seed has no place in any system of scientific agriculture, least of all in dry-farming, where success depends upon the degree with which all conditions are controlled. In all good dry-farm practice seeds should be placed in rows, preferably by means of one of the numerous forms of drill seeders found upon the market. The advantages of the drill are almost self-evident. It permits uniform distribution of the seed, which is indispensable for success on soils that receive a limited rainfall. The seed may be placed at an even depth, which is very necessary, especially in fall sowing, where the seed depends for proper germination upon the moisture already stored in the soil. The deep seeding often necessary under dry-farm conditions makes the drill indispensable. Moreover, Hunt has explained that the drill furrows themselves have definite advantages. During the winter the furrows catch the snow, and because of the protection thus rendered, the seed is less likely to be heaved out by repeated freezing and thawing. The drill furrow also protects to a certain extent against the drying action of winds and in that way, though the furrows are small, they aid materially in enabling the young plant to pass through the winter successfully. The rains of fall and spring are accumulated in the furrows and made easily accessible to plants. Moreover, many of the drills have attachments whereby the soil is pressed around the seed and the topsoil afterwards stirred to prevent evaporation. This permits of a much more rapid and complete germination. The drill, the advantages of which were taught two hundred years ago by Jethro Tull, is one of the most valuable implements of modern agriculture. On dry-farms it is indispensable. The dry-farmer should make a careful study of the drills on the market and choose such as comply with the principles of the successful prosecution of dry-farming. Drill culture is the only method of sowing that can be permitted if uniform success is desired.

The care of the crop

Excepting the special treatment for soil-moisture conservation, dry-farm crops should receive the treatment usually given crops growing under humid conditions. The light rains that frequently fall in autumn sometimes form a crust on the top of the soil, which hinders the proper germination and growth of the fall-sown crop. It may be necessary, therefore, for the farmer to go over the land in the fall with a disk or more preferably with a corrugated roller. Ordinarily, however, after fall sowing there is no further need of treatment until the following spring. The spring treatment is of considerably more importance, for when the warmth of spring and early summer begins to make itself felt, a crust forms over many kinds of dry-farm soils. This is especially true where the soil is of the distinctively arid kind and poor in organic matter. Such a crust should be broken early in order to give the young plants a chance to develop freely. This may be accomplished, as above stated, by the use of a disk, corrugated roller, or ordinary smoothing harrow.

When the young grain is well under way, it may be found to be too thick. If so, the crop maybe thinned by going over the field with a good iron-tooth harrow with the teeth so set as to tear out a portion of the plants. This treatment may enable the remaining plans to mature with the limited amount of moisture in the soil. Paradoxically, if the crop seems to be too thin in the spring, harrowing may also be of service. In such a case the teeth should be slanted backwards and the harrowing done simply for the purpose of stirring the soil without injury to the plant, to conserve the moisture stored in the soil and to accelerate the formation of nitrates. The conserved moisture and added fertility will strengthen the growth and diminish the water requirements of the plants, and thus yield a larger crop. The iron-tooth harrow is a very useful implement on the dry-farm when the crops are young. After the plants are up so high that the harrow cannot be used on them no special care need be given them, unless indeed they are cultivated crops like corn or potatoes which, of course, as explained in previous chapters, should receive continual cultivation.

Harvesting

The methods of harvesting crops on dry-farms are practically those for farms in humid districts. The one great exception may be the use of the header on the grain farms of the dry-farm sections. The header has now become well-nigh general in its use. Instead of cutting and binding the grain, as in the old method, the heads are simply cut off and piled in large stacks which later are threshed. The high straw which later are threshed. The high straw which remains is plowed under in the fall and helps to supply the soil with organic matter. The maintenance of dry-farms for over a generation without the addition of manures has been made possible by the organic matter added to the soil in the decay of the high vigorous straw remaining after the header. In fact, the changes occurring in the soil in connection with the decaying of the header stubble appear to have actually increased the available fertility. Hundreds of Utah dry wheat farms during the last ten or twelve years have increased in fertility, or at least in productive power, due undoubtedly to the introduction of the header system of harvesting. This system of harvesting also makes the practice of fallowing much more effective, for it helps maintain the organic matter which is drawn upon by the fallow season. The header should be used wherever practicable. The fear has been expressed that the high header straw plowed under will make the soil so loose as to render proper sowing difficult and also, because of the easy circulation of air in the upper soil layers, cause a large loss of soil-moisture. This fear has been found to be groundless, for wherever the header straw has been plowed under, especially in connection with fallowing, the soil has been benefited.

Rapidity and economy in harvesting are vital factors in dry-farming, and new devices are constantly being offered to expedite the work. Of recent years the combined harvester and thresher has come into general use. It is a large header combined with an ordinary threshing machine. The grain is headed and threshed in one operation and the sacks dropped along the path of the machine. The straw is scattered over the field where it belongs.

All in all, the question of sowing, care of crop and harvesting may be answered by the methods that have been so well developed in countries of abundant rainfall, except as new methods may be required to offset the deficiency in the rainfall which is the determining condition of dry-farming.

CHAPTER XII

CROPS FOR DRY-FARMING

The work of the fry-farmer is only half done when the soil has been properly prepared, by deep plowing, cultivation, and fallowing, for the planting of the crop. The choice of the choice of the crop, its proper seeding, and its correct care and harvesting are as important as rational soil treatment in the successful pursuit of dry-farming. It is true that in general the kinds of crops ordinarily cultivated in humid regions are grown also on arid lands, but varieties especially adapted to the prevailing dry-farm conditions must be used if any certainty of harvest is desired. Plants possess a marvelous power of adaptation to environment, and this power becomes stronger as successive generations of plants are grown under the given conditions. Thus, plants which have been grown for long periods of time in countries of abundant rainfall and characteristic humid climate and soil yield well under such conditions, but usually suffer and die or at best yield scantily if planted in hot rainless countries with deep soils. Yet, such plants, if grown year after year under arid conditions, become accustomed to warmth and dryness and in time will yield perhaps nearly as well or it may be better in their new surroundings. The dry-farmer who looks for large harvests must use every care to secure varieties of crops that through generations of breeding have become adapted to the conditions prevailing on his farm. home-grown seeds, it grown prevailing on his farm. home-grown seeds, if grown properly, are therefore of the highest value. In fact, in the districts where dry-farming has been practiced longest the best yielding varieties are, with very few exceptions, those that have been grown for many successive years on the same lands. The comparative newness of the attempts to produce profitable crops in the present dry-farming territory and the consequent absence of home-grown seed has rendered it wise to explore other regions of the world, with similar climate conditions, but long inhabited, for suitable crop varieties. The United States Department of Agriculture has accomplished much good work in this direction. The breeding of new varieties by scientific methods is also important, though really valuable results cannot be expected for many years to come. When results do come from breeding experiments, they will probably be of the greatest value to the dry-farmer. Meanwhile, it must be acknowledged that at the present, our knowledge of dry-farm crops is extremely limited. Every year will probably bring new additions to the list and great improvements of the crops and varieties now recommended. The progressive dry-farmer should therefore keep in close touch with state and government workers concerning the best varieties to use

Moreover, while the various sections of the dry-farming territory are alike in receiving a small amount of rainfall, they are widely different in other conditions affecting plant growth, such as soils, winds, average temperature, and character and severity of the winters. Until trials have been made in all these varying localities, it is not safe to make unqualified recommendations of any crop or crop variety. At the present we can only say that for dry-farm purposes we must have plants that will produce the maximum quantity of dry matter with the minimum quantity of water; and that their periods of growth must be the shortest possible. However, enough work has been done to establish some general rules for the guidance of the dry-farmer in the selection of crops. Undoubtedly, we have as yet had only a glimpse of the vast crop possibilities of the dry-farming territory in the United States, as well as in other countries.

Wheat

Wheat is the leading dry-farm crop. Every prospect indicates that it will retain its preeminence. Not only is it the most generally used cereal, but the world is rapidly learning to depend more and more upon the dry-farming areas of the world is rapidly learning to depend more and more upon the dry-farming areas of the would for wheat production. In the arid and semiarid regions it is now a commonly accepted doctrine that upon the expensive irrigated lands should be grown fruits, vegetables, sugar beets, and other intensive crops, while wheat, corn, and other grains and even much or the forage should be grown as extensive crops upon the non-irrigated or dry-farm lands. it is to be hoped that the time is near at hand when it will be a rarity to see grain grown upon irrigated soil, providing the climatic conditions permit the raising of more extensive crops.

In view of the present and future greatness of the wheat crop on semiarid lands, it is very important to secure the varieties that will best meet the varying dry-farm conditions. Much has been done to this end, but more needs to done. Our knowledge of the best wheat is still fragmentary. This is even more true of other dry-farm crops. According to jar dine, the dry-farm wheats grown at present in the United States may be classified as follows:-

1. Hard spring wheats:

   (a) Common

   (b) Durum

2. Winter wheats:

   (a) Hard wheats (Crimean)

   (b) Semihard wheats (Intermountain)

   (c) Soft wheats (Pacific)

The common varieties of hard spring wheats are grown principally in districts where winter wheats have not as yet been successful; that is, in the Dakotas, northwestern Nebraska, and other localities with long winters and periods of alternate thawing and severe freezing. The superior value of winter wheat has been so clearly demonstrated that attempts are being made to develop in every locality winter wheats that can endure the prevailing climatic conditions. Spring wheats are also grown in a scattering way and in small quantities over the whole dry-farm territory. The two most valuable varieties of the common hard spring wheat are Blue Stem and Red Fife, both well-established varieties of excellent milling qualities, grown in immense quantities in the Northeastern corner of the dry-farm territory of the United States and commanding the best prices on the markets of the world. It is notables prices on the markets of the world. It is notable that Red Fife originated in Russia, the country which has given us so many good dry-farm crops.

The durum wheats or macaroni wheats, as they are often called, are also spring wheats, which promise to displace all other spring varieties because of their excellent yields under extreme dry-farm conditions. These wheats, though known for more than a generation through occasional shipments from Russia, Algeria, and Chile, were introduced to the farmers of the United States only in 1900, through the explorations and enthusiastic advocacy of Carleton of the United States Department of Agriculture. Since that time they have been grown in nearly all the dry farm states and especially ion the Great Plains area.

Wherever tried they have yielded well, in some cases as much as the old established winter verities.

The extreme hardness of these wheats made it difficult to induce the millers operating mills fitted for grinding softer wheats to accept them for flour making purposes. This prejudice has, however, gradually vanished, and to day the durum wheats are in great demand, especially for blending with the softer wheats and for the making of macaroni. Recently the popularity of the durum wheats among the farmers has been enhanced, owing to the discovery that they are strongly rust resistant. (See Fig. 31)

The *winter wheats*, as has been repeatedly suggested in preceding chapters, are most desirable or dry-farm purposes, wherever they can be grown, and especially in localities where a fair precipitation occurs in the winter and spring. The hard winter wheats are represented mainly by the Crimean group the chief members of which are Turkey, Krakow, and Crimean. These wheats also originated in Russia and are said to have been brought to the United States a generation ago by Mennonite colonists. At present these whets are grown chiefly in the central and southern parts of the Great Plains area and in Canada, though they are rapidly spreading over the intermountain country. These are good milling wheats of high gluten content and yielding abundantly under dry-farm conditions. It is quite clear that these wheats will soon displace the older winter wheats formerly grown on dry-farms. Turkey wheat promises to become the leading dry-farm wheat. (see Fig. 32)

The semisoft winter wheats are grown chiefly in the intermountain country. They are represented by a very large number of varieties, all tending toward softness and starchiness. This may in part be due to climatic, soil, and irrigation conditions, but is more result of inherent qualities in the varieties used. They are rapidly being displaced by the hard varieties.

The group of soft winter wheats includes numerous varieties grown extensively in the famous wheat districts of California, Oregon, Washington, and northern Idaho. The main varieties are Red Russian and papoose Blue Stem, in Washington and Idaho; Red Chaff and Foise in Oregon, and Defiance, Little Club, Sonar, and White Australian in California. These are all soft, white, and rather poor in gluten. It is believed that under given climatic, soil, and cultural conditions, all wheat varieties will approach one type, distinctive of the conditions in question, and that the California wheat type is a result of prevailing unchangeable conditions. More research is needed, however, before definite principles can be laid down concerning the formation of distinctive wheat types in the various dry-farm sections. Under any condition, a change of seed, keeping improvement always in view, should be beneficial.

Jardine has reminded the dry-farmers of the United States that before the production of wheat on the dry-farms can reach its full possibilities under any acreage, sufficient quantitative must be grown of a few varieties to affect there large markets. This is especially important in the intermountain country where no uniformity exists, but the warning should be heeded also by the pacific coast and Great Plains heat areas. As soon as the best varieties are found they should displace the miscellaneous collection of wheat varieties now grown. The individual farmer can be a law unto himself no more in wheat growing than in fruit growing, if he desires to reap the largest reward of his efforts. Only by uniformity of kind and quality and large production will any one locality impress itself upon the markets and create a demand. The changes now in progress by the dry-farmers of the United States indicate that this lesson has been taken to hearty. The principle is equally important for all countries where dry-farming is practiced.

Other Small grains

Oats is undoubtedly a coming dry-farm crop. Several varieties have been found which yield well on lands that receive and average annual rainfall of less than fifteen inches. Others will no doubt be discovered or developed as special attention is given to dry-farm ousts. Oats occurs as spring and winter verities, but only one winter variety has as yet found place into he list of dry-farm crops. The leading spring varieties of oats are the Sixty Day, Kherson, Burt, and Swedish Select. The one winter variety, which is grown chiefly in Utah, is the Boswell, a black variety originally borough from England about 1901.

Barley, like the other common grains, occurs in charities that grow well on dry-farms. in comparison with wheat very little search has been made for dry-farm barleys, and, naturally, the list of tested varieties is very small. Like wheat and oats, barley occurs in spring and winter varieties, but as in the case of oats only one winter variety has as yet found its way into the approved list of dry-farm crops. The best dry-farm spring barleys are those belonging to the beardless and hull-less types though the more common varieties also yield well, especially the six rowed beardless barley. The winter variety is the Tennessee Winter, which is already well distributed over the Great Plains district.

Rye is one of the surest dry-farm crops. It yields good crops of straw and grain, both of which are valuable stock foods. In fact, the great power of rye to survive and grow luxuriantly under the most trying dry-farm conditions is the chief objection to it. Once started, it is hard to eradicate. Properly cultivated and used either as a stock feed or as green manure, it is very valuable. Rye occurs as both spring and winter verities. The winter varieties are usually most satisfactory.

Carleton has recommended Emmer as a crop peculiarly adapted to semiarid conditions. Emmer is a species of wheat to the berries of which the chaff adheres very closely. It is highly prized as a stock feed. In Russia and Germany it is grown in very large quantities. It is especially adapted to arid and semiarid conditions, at will probably thrive vest where the winters are dry and summers wet. it exists as spring and winter varieties. As with the other small grains, the success of emmer will depend largely upon the satisfactory development of winter varieties.

Corn

Of all crops yet tried on dry-farms, corn is perhaps the most uniformly successful under extreme dry conditions. If the soil treatment and plating have been right, the failures that have been reported may invariably be traced to the use of seed which had not been acclimated. The American Indians grow corn which is excellent for dry-farm purposes; many of the western farmers have likewise produced strains that use the minimum of moisture, and, moreover, corn brought from humid sections adapts itself to arid conditions in a very few years. Escobar reports a native corn frown in Mexico with low stalks and small ears that well endures desert conditions. In extremely dry years corn does not always produce a profitable crop of seed, but the crop as a whole, for forage purposes, seldom fails to pay expenses and leave a margin for profit. In wetter years there is a corresponding increase of the corn crop. The dry-farming territory does not yet realize the value of corn as a dry-farm crop. The known facts concerning corn make it safe to predict, however, that its dry farm acreage will increase rapidly, and that in time it will Crowe the wheat crop for preeminence.

Sorghums

Among dry-fames crops not popularly known are the sorghums, which promise to become excellent yielder under arid conditions. The sorghums are supposed to have come from the tropical sections of the globe, but they are now scattered over the earth in all climes.

The sorghums have been known in the United States for over half a century, but it was only when dry-farming began to develop so tremendously that the drought-resisting power of the sorghums was recalled. According to Ball, the sorghums fall into the following classes:-

THE SORGHUMS

1. Broom corns
2. Sorgas or sweet sorghums
3. Kafirs
4. Durras

The broom corns are grown only for their rush, and are not considered in dry-farming; the sorgas for forage and syrups, and are especially adapted for irrigation or humid conditions, though they are said to endure dry-farm conditions better than corn. The Kafirs are dry-farm crops and are grown for grain and forage. This group includes Red Kafirs, White Kafirs, Black-hulled White Kafirs, and White Milo, all of which are valuable for dry-farming. The Durras are grown almost exclusively for seed and include Jerusalem corn, Brown Durra, and Milo. The work of Ball has made Milo one of the most important dry-farm crops. As improved, the crop is from four to four and a half feet high, with mostly erect heads, carrying a large quantity if seeds. Milo is already a staple crop in parts of Texas, Oklahoma, Kansas, and New Mexico. It has father been shown to be adapted to conditions in the Dakotas, Nebraska, Colorado, Arizona, Utah, and Idaho. It will probably be found, in some parietal form, valuable over the whole dry-farms territory where the altitude is not too high and the average temperature not too low. it has yielded an average of forty bushels of seed to the acre.

Lucern or alfalfa

Next to human intelligence and industry, alfalfa has probably been chief factor in the development of the irrigated west. It has made possible a rational system of agricultural with the live-stock industry and the maintenance of soil fertility as the central considerations. Alfalfa is now being recognized as a desirable crop in humid as well as in irrigated sections, and it is probable that alfalfa will soon become the chief hay crop of the United States. Originally, Lucern cane from the hot dry countries of Asia, Where it supplied feed to the animals of the first historical peoples. Moreover, its long taps roots, penetrating sometimes forty or fifty feet into the ground. Suggest that Lucern may make ready use of deeply stored soil moisture. On these considerations, alone, Lucern should prove itself a crop well suited for dry-farming. In fact, it has been demonstrated that where conditions are favorable, Lucern may be made to yield profitable crops under a rainfall between twelve and fifteen inches. Alfalfa prefers calcareous loamy soils; sandy and heavy clay soils are not so well adapted for successful alfalfa production. Under dry-farm conditions the utmost car must be used to prevent too thick seeding. The vast majority of alfalfa failures on dry-farms have resulted from an insufficient supply of moisture for the thickly planted crop. The alfalfa field does not attain its maturity until after the second year, and a crop which looks just right the second year will probably be much too thick the third and fourth years. From four to six pound of seed per acre are usually ample. Another main cause of failure is the common idea that the Lucern field needs little or no cultivation, when, in fact, the alfalfa field should receive as careful soil treatment as the wheat field. Heavy, thought disking in spring or fall, or both, is advisable, for it leaves the topsoil in a condition to prevent evaporation and admit air. In Asiatic and North African countries, Lucern is frequently cultivated between rows throughout the hot season. This had been tried by Brand in this country and with very good results. Since the crop sho8uld always be sown with a drill, it is comparatively easy to regulate the distance between the row\s so that cultivating implements may be used. If thin seeding and thorough soil stirring are practiced, Lucern usually grows well, and with such treatment should become one of the great dry-farm crops. The yield of hay is not large, but sufficient to leave a comfortable margin of profit. Many farmers find it more profitable to grow dry-farm Lucern for seed. In good years from fifty to onc hindered and fifty dollars may be taken from an acre of Lucern seed. However, at the present, the principles of Lucern seed production are not well established, and the seed crop is uncertain.

Alfalfa is leguminous crop and gathers nitrogen from the air. It is therefore a food fertilizer. The question of soil fertility well become more important with the passing of the years, and the value of Lucern as a land improver will then be more evident than it is to-day

Other leguminous crops

The groups of leguminous or pod-bearing crops is of great importance; first, because it is rich in nitrogenous substances which are valuable animal foods, and, secondly, because it has the power of gathering nitrogen from the air, which can be used for maintaining the fertility of the soil. Dry-farming will not be a wholly safe practice of agriculture until suitable leguminous crops are found and made part of the crop system. It is notable that over the whole of the dry-farm territory of this and other countries wild legume found plants flourish. That is, nitrogen-gathering plants are at work on the deserts. The farmer upsets this natural order of things by cropping the land with wheat and wheat only, so long as the land will produce profitable. The leguminous plants native to dry-farm areas have not ads yet been subjected to extensive economic study, and in trusty very little is known concerning leguminous plant s adapted to dry-farming.

In California, Colorado, and other dry-farm states the field pea has been frown with great profit. Indeed it has been found much more profitable than wheat production. The field bean, likewise, has been grown successfully under dry-farm conditions, under a great variety of climates. In Mexico and other southern climates, the native population produce large quantities of beans upon their dry lands. Shaw suggests that sanfoin, long famous for its service to European agriculture may be found to be a profitable dry-farm crop, and that sand vetch promises to become an excellent dry-farm crop. It is fiery likely, however, that many of the leguminous crops which have been developed under conditions of abundant rainfall will be valueless on dry-farm lands. Every year will furnish new and more complete information on this subject. Leguminous plants will surely become important members of the association of dry-farm crops.

Tree and shrubs

So far, trees cannot be said to be dry-farm crops, though facts are on record that indicate that by the application of coerce t dry-farm principles trees may be made to grow and yield profitable on dry-farm lands. Of course, it is a well-known fact that native trees of various kinds are occasionally found growing on the deserts, where the rainfall is very light and the soil has been given no care. Examples of such vegetation are the native cedars found throughout the Great Basic region and the mesquite tree in Arizona and the Southwest. Few farmers in the arid region have as yet undertaken tree culture without the aid of irrigation.

At least one peach orchard is known in Utah which grown under a rainfall of about fifteen inches without irrigation and produces regularly a small crop of most delicious fruit. parsons describes his Colorado dry farm orchard in which, under a rainfall of about fourteen inches, he frowns, with great profit, cherries, plums, and apples. A number of prospering young orchards are growing without irrigation in the Great Plains area. Mason discovered a few years ago two olive orchards in Arizona and the Colorado desert which, planted about fourteen years previously, were thriving under an annual rainfall of eight and a half and four and a half inches, respectively. Theses olive orchards in Arizona and the Colorado desert which, planted about fourteen years previously, were thriving under an annual rainfall of eight and a half and four and a half inches, respectively. These olive orchards had been set out under canals which later failed. Such attested facts lead to the thought that trees may yet take their place as dry-farm crops. This hope is strengthened when it is recalled that the great nations of antiquity, living in countries of low rainfall, grew profitably and without irrigation many valuable trees, some of which are still cultivated in those countries. The olive industry, for example, is even now being successfully developed by modern methods in Asiatic and African sections, where the average annual rainfall is under ten inches. Since 1881, under French management, the dry-farm olive trees around Tunis have increased from 45,000 to suggest as trees that do well in the arid parts of the old world the so-called "Chinese date" or Jujube tree, the sycamore fig, and the Carob tree, which yields the "ST. John's Bread" so dear to childhood. Of this last tree Aaronsohn says that twenty trees to the acre, under a rainfall of twelve inches, will produce 8000 pounds of fruit containing 40 per cent of sugar and 7 to 8 per cent of protein. This surpasses the best harvest of alfalfa. Kearney, who has made a special study of dry-land olive culture in northern Africa, states that in his belief a large variety of fruit trees may be found which will do well under arid and semiarid conditions, and may even yield more profit than the grains.

It is also said that many shade and ornamental and other useful plants can be grown on dry-farms; as, for instance, locust, elm, black walnut, silver poplar, catalpa, live oak, black oak, yellow pine, red source, Douglas fir, and cedar.

The secret of success in tree growing n dry-farms seems to lie, first, in plating a few trees per acre, the distance apart should be twice the ordinary distance, and, secondly, in applying vigorously and unceasingly the established principles of soil cultivation. In a soil stored deeply with moisture and properly cultivated, most plants will grow. If the soil has not been carefully fallowed before planting, it may be necessary water the young trees slightly during the first two seasons.

Small fruits have been tried on many farms with great success. Plums, currants and gooseberries have all been successful. Grapes grow and yield well in many dry-farm districts, especially along the warm foothills of the Great Basin.

Tree growing on dry-farm lands is not yet well established and, therefore, should be undertaken with great care. Varieties accustomed to the climatic environment should be chosen, and the principles outlined in the preceding pages should be carefully used.

Potatoes

In recent years, potatoes have become one of the best dry-farm crops. Almost wherever tried on lands under a rainfall of twelve inches or more potatoes have given comparatively large yields. To-day, the growing of dry-farm potatoes is becoming an important industry. The principles of light seeding and thorough cultivation are indispensable for success. Potatoes are well adapted for use in rotations, where summer fallowing is not thought desirable. Macdonald enumerates the following as the best varieties at present used on dry-farms: Ohio, Mammoth, Pearl, Rural New Yorker, and Burbank.

Miscellaneous

A further list of dry-farm crops would include representatives of nearly all economic plants, most of them tried in small quantity in various localities. Sugar beets, vegetables, bulbous plants, etc., have all been grown without irrigation under dry-farm conditions. Some of these will no doubt be found to be and will then be brought into the commercial scheme of dry-farming.

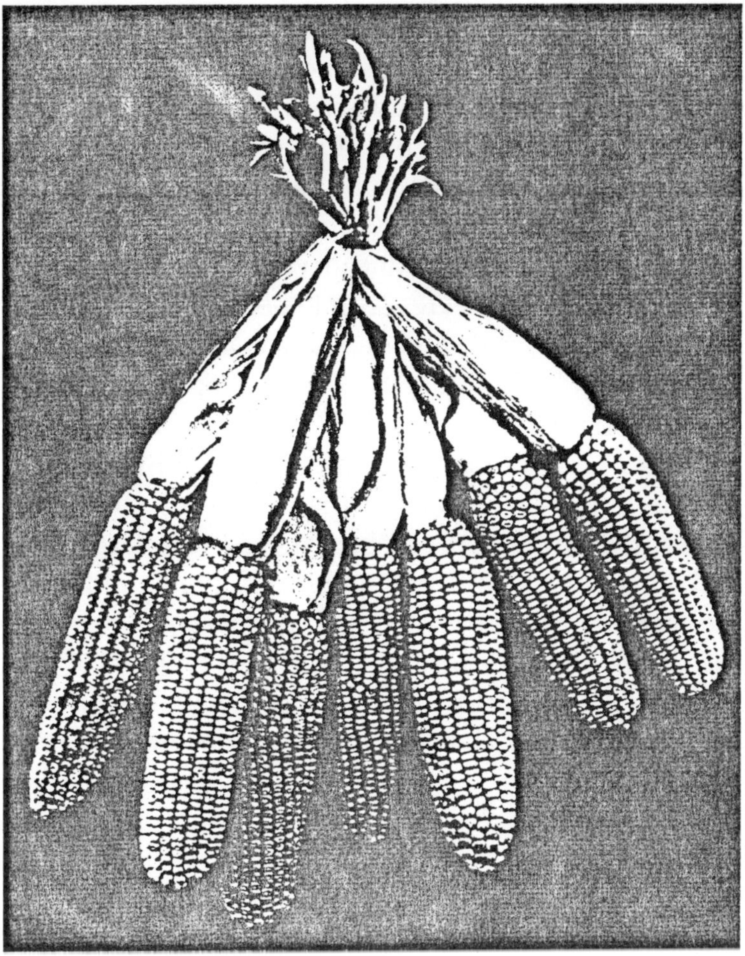

Fig. 30: Ears of dry-Farm corn Montana

Meanwhile, the crop problems of dry-farming demand that much careful work be done in the immediate future by the agencies having such work in charge. The best varieties of crops already in profitable use need to be determined. More new plants from all parts of the world need to be brought to this new dry-farm territory and tried out. Many of the native plants need examination with a view to their economic use. For instance, the sag lily builds, upon which the Utah pioneers subsisted for several seasons of famine, may possible by made a cultivated crop. Finally, it remains to be said that it is doubtful wisdom to attempt to grow the more intensive crops on dry-farms. Irrigation and dry-farming will always go together. They are supplementary systems of agriculture in arid and semiarid regions. On the irrigated lands should be frown the crops that require much labor per acre and that in return yield largely per acre. New crops and varieties should be sought for the irrigated farms. On the dry-farms should be grown the crops that can be handled in a large way and at a small cost per acre, and that yield only moderate acre returns. By such cooperation between irrigation and dry-farming will the regions of the world with a scanty rainfall become the healthiest, wealthiest, happiest, and most populous on earth.

CHAPTER XIII

THE COMPOSITION OF DRY-FARM CROPS

The acre-yields of crops on dry-farms, even under the most favorable methods of culture, are likely to be much smaller than in humid sections with fertile soils. The necessity for frequent fallowing or resting periods over a large portion of the dry-farm territory further decreases the average annual yield. It does not follow from than humid- or irrigation-farming, for it has been fully demonstrated that the profit on the investment is as high under proper dry-farming as under any other similar generally adopted system of farming in any part of the world. Yet the practice of dry-farming would appear to be, and indeed would be, much more desirable could the crop yield be increased. The discovery of any condition which will offset the small annual yields is, condition which will offset the small annual yields is, therefore, of the highest importance to the advancement of dry-farming. The recognition of the superior quality of practically all crops grown without irrigation under a limited rainfall has done much to stimulate faith in the great profitableness of dry-farming.

Fig. 31: heads of macaroni wheat.1 Kubanka 2, Nicaragua 3, velvet don 4 Black Don 5, Wild Goose. These among the best among the best drought-resistant spring wheats.

As the varying nature of the materials used by man for food, clothing and shelter has become more clearly under stood, more attention has been given to the valuation of commercial products on the basis of quality as well as of quantity. Sugar beets, for instance, are bought by the sugar content; and many factories of Europe vary the price paid according to the sugar contained by the beets, the millers, especially in certain parts of the country where wheat has deteriorated distinguish carefully between the flour-producing qualities of wheats from various sections and fix the price concerning the real nutritive value of various foods is being sought eagerly, and foods known to possess the highest value in the maintenance of life are displacing, even at a higher cost, the inferior products. The quality valuation is, in fact being extended as rapidly as the growth of knowledge will permit tot eh chief food materials of commerce. As this practice becomes fixed the dry-farmer will his products, for it is undoubtedly true that from the point of vie of quality, partition with any farm products on the markets of the world.

Proportion of plant parts

It need hardly be said, after the discussions in the preceding chapters, that the nature of plant growth is deeply modified by the arid conditions prevailing in dry-farming. This shows itself first in the proportion of the various plant parts, such as roots stems, leaves, and seeds. The root systems of dry-farm crops are generally greatly developed and it is a common observation that in adverse seasons the plants that possess the largest and most vigorous roots endure best the drought and burning heat. The first function of the leaves is to gather materials for the building and strengthening of the roots, and only after this has been done do the stems lengthen and the leaves thicken. Usually, the short season is largely gone before the stem and leaf growth beings, and consequently, a somewhat dwarfed appearance is characteristic of dry-farms crops. The size of sugar beets, potato tubers, and such underground parts depends upon the available water and food supply when the plant has established a satisfactory root and leaf system. If the water and food are scarce, a thin beet results; if abundant, a well-filled beet may result.

Dry-farming is characterized by a somewhat short season. Even if good growing weather prevails, the decrease of water in the soil has the effect of hastening maturity. The formation of flowers and seed beings, therefore, earlier and is completed more quickly under arid than under humid conditions. Moreover, and resulting probably from the greater abundance of material stored in the root system. The proportion of hands to leaves and stems is highest in dry-farm crops. In fact, if is a general law that the proportion of heads to straw in grain crops increases as the water simply decreases. This is shown very well even under humid or irrigation conditions when different seasons or different applications of irrigation water are compared. For instance, Hall quotes from the Rothamusted experiments to the effect that in 1879, which was a wet year (41 inches), the wheat crop yielded 38 pounds of grain for every 100 pounds of straw; whereas, in 1893 which was a dry year (23 inches), the whet crop yielded 95 pounds of grain to every 100 pounds of straw. The Utah station likewise has established the same law under arid conditions. In one series of experiments it was shown as an average of three years' trial that a field which had received 22.5 inches of irrigation water produced a wheat crop that gave 67 pounds of grain to every 100 pounds of straw; while another field which received only 7.5 inches of irrigation water produced a crop that gave 100 pounds of grain for every 100 pounds of straw. Since wheat is grown essentially for the grain, such a variation is of tremendous importance. The amount of available water affects every part of the plant. Thus, as an illustration, Carleton states that the per cent of meat in oats frown in Wisconsin under humid conditions was 67.24, while in North Dakota, Kansas, and Montana, under arid and semiarid condition, it was 71.51. Similar variations of plants may be observed as a direct result of varying the amount of available water. In general, then, it may be said that the roots of dry-farm crops ware well developed; the parts above ground somewhat dwarfed; the proportion of seed to straw high, and the proportion of meat or nutritive materials in the plant parts likewise high.

The water in dry-farm crops

One of the constant constituents of all plants and plant parts is water. Hay, flour, and starch contain comparatively large quantities of water, which can be removed only by ahead. The water in green plants is often very large. In young lucre, for instance, it reaches 85 per cent, and in young peas nearly 90 per cent, or more than is found is good cow's milk. The water so held by plants has no nutritive value above ordinary water. It is, therefore, profitable for the consumer to buy dry foods. In this particular, again, dry-farm crops have a distinct advantage. During growth there is not perhaps a great difference in the water content of plants, due t climatic differences, but after harvest the drying-out process goes on much more completely in dry-farm than in humid districts. Hay, cured in humid regions, often contains from12 to 20 per cent of water; in arid climates if contains as little as 54 per cent and seldom more than 12 per cent. The drier hay is naturally more valuable pound than the moister hay, and a difference in pricc, based upon the difference in water content, is already being felt in certain sections of the West.

The moisture content of dry-farm wheat, the chief dry-farm crop, is even more important. According to Wiley the average water content of wheat for the United States is 10.62 per cent, ranging from 15 to 7 per cent. Stewart and Greaves examined a large number of wheats grown on the dry-farms of Utah and found that the average per cent of water in the common bread varieties was 8.46 and in the durum varieties 8.89. This means that the Utah dry-farm wheats transported to ordinary humid conditions would take up enough water from the air to increase their weight one fortieth, or 2½ per cent, before they reached the average water content of American wheats. In other words 1,000,000 bushels of Utah dry-farm wheat contains much nutritive matter as 1,025,000 bushels of much grown and kept under humid conditions. This difference should be and now is recognized in the prices paid. In fact, shrewd dealers, acquainted with the dryness of dry-farm wheat, have for some years bought wheat from the dry-farms at a slightly increased price, and trusted to the increase in weight due to water absorption in more humid climates for their profits. The time should be near at hand when grains and similar products should been purchased upon basis of a moisture test.

While it is undoubtedly true that dry-farm crops are naturally drier than those of humid countries, yet it must also be kept in mind that the driest dry farm crops are always obtained where the summers are hot and rainless. In sections where the precipitation comes chiefly in the spring and summer the difference would not be so great. Therefore, the crops raised on the Great Plains would not be so dry as those raised in California or in the Great Basin. Yet, wherever the annual rainfall is so small as to establish dry-farm conditions, whether it comes in the winter or summer, the cured crops are drier than those produced under conditions of a much higher rainfall, and dry farmers should insist that, so far as possible in the future, sales be based on dry matter.

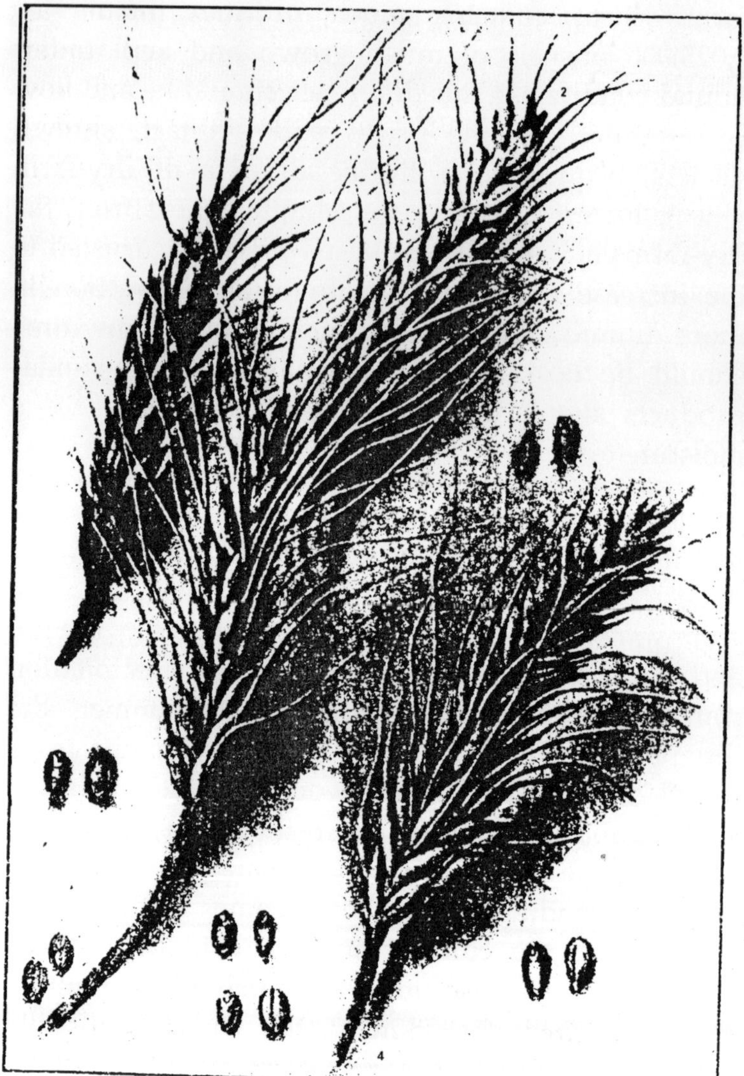

Fig. 32: Heads of hard winter wheats. 1, Turkey (Crimean). 2, Odessa white chaff. 3 Odessa Red Chaff. 4, Rumanian white Chaff 5, Krakow 6 Ulta.

The nutritive substances in crops

The dry matter of all plants and plant parts consists of three very distinct classes of substances: First, ash or the mineral constituents. Ash is used by the body in building bones and in supplying the blood with compounds essential to the various life processes. Second, protein or the substances containing the element nitrogen. Proton is used by the body in making blood, muscle, tendons, hair, and nails, and under certain conditions it is burned within the body for the production of heat. Protein is perhaps the most important food constituent. Third, non-nitrogenous substances, including fats, woody fiber, and nitrogen-free extract, a name given to the group of sugars, starches, and related substances. These substances are used by the body in the production of fact, and large also burned for the production of heat. Of these valuable food constituents protein is probably the most important, first, because it forms the most important tissues of the body and, secondly, because it is less abundant than the fats, starches, and sugars. Indeed, plants rich in protein nearly always command the highest prices.

The composition of any class of plants varies considerably indifferent localities and in different seasons. This may be due to the nature of the soil, or to the fertilizer applied. Though variations in plant composition resulting from soil conditions are comparatively small. The greater variations are almost wholly the result of carrying climate and water supply. As far as it is now known the strongest single factor enchaining the composition of plants is the amount of water available to the growing plant.

Variations due to varying water supply

The Utah station has conducted numerous experiments upon the effect of water upon plant composition. The method in every case has been to apply different amounts of water throughout the growing season on contiguous plats of uniform land. Some of the early results are shown by the following table:-

TABLE: 18
THE EFFECT OF WATER ON THE PERCENTAGE COMPOSITION OF PLANT PARTS

Inches of Water applied	Ash	Protein	Fat	Fiber	Nitrogen free extract
		Corn Kernels			
7.5	1.62	15.08	6.02	1.89	75.39
15.0	1.65	13.48	6.16	1.91	76.86
37.3	1.62	12.52	6.26	1.89	77.72
		Oat Kernels			
7.0	3.26	20.79	3.91	9.02	63.02
13.2	4.52	17.29	4.19	10.76	63.25
30.0	4.49	15.49	4.59	10.92	64.51
		Wheat Kernels			
4.6	2.70	26.72	2.37	5.44	62.77
10.3	2.54	19.93	2.09	4.47	70.97
21.1	2.50	16.99	1.97	3.97	74.62
		Pea Kernels			
7.5	1.17	31.16	1.70	7.88	58.09

Inches of Water applied	Ash	Protein	Fat	Fiber	Nitrogen free extract
15.0	2.76	28.37	0.87	7.14	60.84
30.5	2.99	21.29	1.16	6.78	67.78
		Potato Tubers			
8.0	6.68	11.83	0.55	2.69	78.25
15.0	4.85	12.52	0.33	2.21	80.08
40.0	4.87	8.33	0.79	2.06	83.95
		Sugar Beets			
12.3	4.76	9.68	0.29	5.37	79.91
21.0	4.98	7.50	0.18	6.02	81.32
40.8	4.69	5.63	0.45	5.68	83.55

Even a casual study of this table shows that the quantity of water used influenced the composition of the plant parts. The ash and the fiber do not appear to be greatly influenced, but the other constituents vary with considerable regularity with the variations in the amount of irrigation water. The protein shows the greatest variation. As the irrigation water is increased, the percentage of protein decreases. In the case of wheat the variation was over 9 per cent. The percentage of fact and nitrogen-free extract, on the other hand, becomes larges as the water increase. That is, crops grown with little water, as in dry-farming, are rich in the important flesh-and blood-forming substance protein, and comparatively poor in fat, sugar, starch, and other of the more abundant heat and fat-producing substances. This difference is of tremendous importance in placing dry-farm products on the food markets of the world. Not only seeds, tubers, and roots show this variation, but the stems and leaves of plants grown with little water are found to contain a higher percentage of protein than those grown in more humid climates.

The direct effect of water upon the composition of plants has been observed by many students. For instance, Mayer, working in Holland, found that, in a soil containing throughout the season 10 per cent of water, oats was produced containing 10.6 per cent of protein; in soil containing 30 per cent of water, the protein percentage was only 5.6 per cent, and in soil containing 70 per cent of water, it was only 5.2 per cent. Carleton, in a study of analyses of the same varieties of wheat grown in humid and semiarid districts of the United States, found that the percentage of protein in wheat from the semiarid area was 14.4 per cent as against 11.94 per cent in the wheat from the humid area. The average protein content of the wheat of the United States is a little more than 12 per cent; Stewart and Greaves found an average of 16.76 per cent of protein in Utah dry-farm wheats of the common bread varieties and 17.14 per cent in the durum varieties. The experiments conducted at Rothamsted, England, as given by Hall, confirm theses results. For examples, during 1893, a very dry year, barley kernels contained 12.99 per cent of protein, while in 1894, a wet, though free growing year, the barley contained only 9.81 per cent of protein. Quotations might be multiplied confirming the principle that crops grown with little water contain much protein and little heat-and fat producing substances.

Climate and composition

The general climate, especially as regards the length of the growing season and naturally including the water supply, has a strong effect upon the composition of plants. Carleton observed that the same varieties of wheat grown art Nephi, Utah, contained 16.61 per cent protein; at Amarillo, Taxes, 15.25 per cent; and at McPherson, Kansas, a humid station, 13.054 per cent. This variation is undoubtedly due in part to the varying annual precipitation but, also, and in large part, to the varying general climatic conditions at the three stations.

An extremely interesting and important experiment, showing the effect of locality upon the composition of wheat kernels, is reported by Lecher and Leavitt. Wheat grown in 1905 in Kansas was planted in 19806 in Kansas, California, and Texas. In 1907 samples of the seeds grown at these three points were planted side by side at each of the three states. All the crops from the three localities were analyzed separately each year. Some of the results of this experiment are shown in the following table:-

The results arc striking and convincing. The original seed grown in Kansas in 1905 contained 16.22 per cent of protcin. The 1906 crop grown from this seed in Kansas contained 19.13 per cent protein. In California, 10.38 percent; and in Taxes, 12.18 percent. In 1907 the crop harvested in Kansas from the 1906 seed from these widely separated place and of very different composition contained uniformly some what more than 22 per cent of protein; harvested in California, somewhat more than 11 per cent; and harvested in Taxes about 18 per cent. In short, the composition of wheat kernels is independent of the composite on of the seed or the nature of the soil, but depends primarily upon the prevailing climatic conditions, including the waters supply.

TABLE: 19
EFFECT OF LOCALITY ON COMPOSITION OF CRIMEAN WHEAT

Determination	Grown in Kansas	Grown in California	Grown in Texas
	Original Seed, Kansas, 1905		
Protein	16.22		
Weight per bushel (ibs.)	56.50		
Flinty (per cent)	98.00		
	1906 Crop from Kansas Seed of 1905		
Protein	19.13	10.38	12.18
Weight per bushel (ibs)	58.8	59.4	58.9
Flinty (per cent)	100.00	36.0	

1907 CROP FROM SEED OF 1906

	From Kansas	From California	From Texas	Kansas	California	Taxes	Kansas	California	Taxes
Portion	22.23	22.23	22.18	11.00	11.33	11.37	16.97	18.22	18.21
Weight per bushel (ibs)	51.3	51.3	50.7	61.3	61.8	62.3	58.5	57.3	58.6
Flinty	100.0	100.0	100.0	50.0	60.0	50.0	58.5	57.3	58.6

The weight of the wheat per bushel, that is, the average size and weight of the wheat kernel, and also the hardness or flinty character of the kernels, were strongly effected by the varying climatic conditions. It is generally true that dry-farm grain weights more per bushel than grain grown under humid conditions; hardness usually accompanies a high protein content and is therefore characteristic of dry-farm wheat. These notable lessons teach the futility of bringing in new seed from far distant places in the hope that better and larger crops may be secured. The conditions under which growth occurs determine chiefly the nature of the crop. It is a common experience in the West that farmers who do not understand this principle send to the Middle West for seed corn, with the result that great crops of stalks and leaves with not ears are obtained. The only safe rule for the dry farmer to follow is tousle seed which has been grown for many years under dry-farm conditions.

A reason for variation in composition

It is possible toughest a reason for the high protein content of dry-farm crops. It is well known that all plants secure most of their nitrogen early in the growing period. From the nitrogen, protein is formed, and all young plants are, therefore, very rich in protein. As the plant becomes older, litter more protein is added, but more and more carbon is taken from the air to form the fats. Starches, sugars, and other non-nitrogenous substances. Consequently, the proportion or parentage of protein becomes smaller as the plant becomes older. The impelling purpose of the plant is to produce seed. Whenever the water supply goings to give out, or the season shortens in any other way, the plant immediately beans to ripen. Now, the essential effect of dry-farm conditions is to shorten the season; the comparatively young plants, yet rich in protein, begin to produce seed; and at harvest, seed, and leaves, and stalks are rich in the flesh-and blood-forming element of plants. In more humid countries plants delay the time of seed production and thus enable the plants to store up more carbon and thus reduce the percent of protein. The short growing season, induced by the shortness of water, is undoubtedly the main reason for the higher protein content and consequently higher nutritive value of all dry-farm crops.

Nutritive value of dry-farm hay, straw, and flour

All the parts of dry-farm crops are highly nutritious. This needs to be more clearly understood by the dry-farmers. Dry-farm hay, for instance, be-cause of its high protein content, may be fed with crops not so rich in this element, thereby making a larger profit for the farmer. Dry-farm straw often has the feeding value of good hay, as has been demonstrated by analyses and by feeding tests conducted in times of hay scarcity. Especially is the header straw of high feeding value, for it represents the upper and more nutritious ends of the stalks. Dry-farm straw, therefore, should be carefully kept and fed to animals instead of being scattered over the ground or even burned as is too often the case. Only few feeding experiments having in view the relative feeding value of dry-farm crops have as yet been made, but the few on record agree in showing the superior value of dry-farm crops, whether fed singly or in combination.

The differences in the chemical composition of plants and plant products induced by differences in the water-supply and climatic environment appear in the manufactured products, such as flour, bran and shorts. Flour made from Fife wheat grown on the dry-farms of Utah contained practically 16 per cent of protein, while flour made from Fife wheat grown in Maine and the Middle West is reported by the Maine Station as containing from 13.03 to 13.75 per cent of protein. Flour made from Blue Stem wheat grown on the Utah dry-farms contained 15.52 per cent of protein; from the same variety grow in Maine and n the Middle West 11.69 and 11.51 per cent of protein respectively. The moist and dry gluten, the goad in and the gluten in, all of which make possible the best and most nourishing kinds of breads, are present in largest quantity and best proportion in flour made from wheats grown under typical dry-farm conditions. The by-products of the milling process, likewise, are rich in nutritive elements.

Future Needs

It has already been pointed out that there is a growing tendency to purchase food material on the basis of composition. New discoveries in the domains of plant composition and animal nutrition and the improved methods of rapid and accurate valuation will accelerate this tendency. Even now, manufactures of food products print on cartons and in advertising matter quality reasons for the superior food values of certain articles. At least one firm produces two parallel sets of its manufactured foods, one for the man who does hard physical labor, and the other for the brain worker. Quality, as related to the needs of the body, whether of beast or man is rapidly becoming the first question in judging any food material. The present era of high prices makes this matter even more important.

In view of this condition and tendency, the fact that dry-farm products are unusually rich in the most valuable nutritive materials is of tremendous importance to the development of dry-farming. The small average yields of dry-farm crops do not look so small when it is known that they command higher prices per pound in competition with the larger crops of more humid climates. More elaborate investigations should be undertaken to determine the quality of crops grown in different dry-farm districts. As far as possible each section, great or small, should confine itself to the growing of a variety of each crop yielding well and possessing the highest nutritive value. In that manner each section of the great dry-farm territory would soon come to stand for some dependable special quality that would compel a first-class market. Further, the superior feeding value of dry-farm products should be thoroughly advertised among the consumers in order the create a demand on the markets for a quality valuation. A few years of such systematic honest work would do must to improve the financial basis of dry-farming.

CHAPTER XIV

MAINTAINING THE SOIL FERTILITY

All plants when carefully burned leave a portion of ash, raging widely in quantity, averaging about 5 per cent, and often exceeding 10 per cent of the dry weight of the pliant. This plant ash represents inorganic substances taken from the soil by the roots. In addition, the nitrogen of plants, averaging about 2 per cent and often amounting to 4 per cent, which, in boring, passes off a gaseous form, is also usually taken from the soil by the plant roots. A comparatively large quantity of the plant is, therefore, drawn directly from the soil. Among the ash ingredients are many which are taken up by the plant simply because they are present in the soil; others, on the other hand, as has been shown by numerous classical investigations, are indispensable to plant growth. If any one of these indispensable ash ingredients be absent, it is impossible for a plant to mature on such a soil. In fact, it is pretty well established that, providing the physical conditions and the water supply are satisfactory, the fertility of a soil depends largely upon the amount of available ash ingredients, or plant-food.

A clear distinction must be made between the total and available plant-food. The essential plant foods often occur in insoluble combinations, valueless to plants only the plant-foods that are soluble in the soil-water or in the juices of plant roots are of value to plants. It is true that practically all soils contain all the indispensable plant-foods; it is also true, however, that inmost soils they are present, as available plant-foods, in comparatively small quantities. When crops are removed from the land year after year, without any return being made, it naturally follows that under ordinary conditions the amount of viable plant-food is diminished, with a strong probability of a corresponding diminution in crop-producing power. In fact, the soils of many of the older countries have even permanently injured by conterminous cropping, with nothing returned, practiced through centuries. Even in many of the younger states, continuous cropping to wheat or other crops for a generation or less ha resulted in a large debase in the crop yield.

Practice and experiment have shown that such demising fertility may be retarded or wholly avoided, first, by so working or cultivating the soil as to set free much of the insoluble pliant-food and, secondly, retiring to the soil all or part of the plant-food taken away. The recent development of the commercial fertile industry is a response to this trust. It may be said that, so far as the agricultural soils of the would are now known, only three of he essential plant-foods are likely to be absent, namely, potash, phosphoric acid, and nitrogen; of these, by far the most important is nitrogen. The whole question of maintaining the supply of plant-foods in the soil concerns itself in the main with the supply of these three substances.

The persistent fertility of dry-farms

This view has been taken because of the well-known faction recent years; numerous farmers and some investigators have stated that under dry-farm conditions the fertility of soils is not impaired by cropping without maturing. That in localities where dry-farming has been practiced on the same soils from twenty-five to forty-five years, without the addition of manures, the average crop yield has not only failed to diminish, but in most cases has increased. In fact, it is the almost unanimous testimony of the oldest dry-farmers of there United States, operating under a rainfall from twelve to twenty inches, that the crop yields have increased as the cultural methods have been perfected, if any adverse effect of the steady removal of plant-foods has occurred, it has been wholly overshadowed by other factors. The older dry-farms in Utah, for instance, which are among the oldest of the country, have never been manure, yet are yielding better. To-day than they did a generation ago, strangely enough, this is not true of the irrigated farms, operating under like soil and climatic conditions. This behavior of crop production under dry-farm conditions has led to the belief that the question of soil fertility is not an important one to dry-farmers. Nevertheless, if our present theories of plant nutrition are correct, it is also true that if continuous cropping is practiced on our dry-farm soils without some form of maturing the time must come when the productive power of the soils will be injured and the only recourse of the rarer will be to return to the soils some of the plan-food taken from it.

The view that silo fertility is not diminished by dry-farming appears at first sight to be strengthened by the results obtained by investigations who have made determinations of the actual plant-food in soils that have long been dry-farmed. The sparsely settled condition of the dry-farmed. The sparsely settled condition of the dry-farm territory furnishes as yet an excellent opportunity to compare virgin and dry-farmed lands and which frequently may be found side by side in even the older dry-farm sections. Stewart found that Utah dry-farm soils, cultivated for fifteen to forty years and never manure, were in many cases richer in nitrogen than engorging virgin lands. Bradley found that the solid of the great dry-farm wheat belt of Eastern Oregon contained, after having been farmed for a quarter of a century, practically as much nitrogen as the adjoining virgin lands. The determinations were made to a depth of eighteen inches, always and Trumbull, on thee other hand, found in a soil from Indian Head, Saskatchewan, that in twenty-five years of cultivation the total amount of nitrogen had been reduced about one third, though the alternation of fallow and crop, commonly practiced in dry-farming, did not show a greater loss of soil nitrogen than other methods of cultivation. It must be kept in mind that the soil of Indian Head

Contains from two to three times as much nitrogen as is ordinarily found in the soils of the Great Plains and from 5three to four times as much as is found in the soils of the Great Basin and the High Plateaus. It may be assumed, therefore, that the Indian Head soil was peculiarly liable to nitrogen losses. Hidden, in an investigation of the nitrogen content of Colorado soils, has come to the conclusion that arid conditions, like those of Colorado, favor the direct accumulation of nitrogen in soils. All in all, the undiminished crop yield and the composition of the cultivated field lead to the belief that soil-fertility problems under dry-farm conditions are widely different from the old well-known problem under humid conditions.

Reasons for dry-farming fertility

It is not really difficult to understand why the yields and, apparently, the fertility of dry-farms have continued to increase during the period of recorded dry-farm history---- nearly half a century.

First, the intrinsic fertility of arid as compared with humid solid is very high. (see Chapter V.) The production and removal of many successive bountiful crops would not have as marked an effect on arid as on humid soils, for both yield and composition change more slowly on ferule soils. The natural extraordinarily high fertility of dry-farm soils explains, therefore, primarily and chiefly, the increasing yields on dry-farm soils that receive proper cultivation.

The intrinsic fertility of arid soils is not alone sufficient to explain the increase in plant-food which undoubtedly occurs in the upper foot or two of cultivated dry-farm lands. In seeking a suitable explanation of this phenomenon it must be recalled that the p4roporation of available plant-food in arid soils is very uniform to great depths, and that plants grown under proper dry-farm conditions are deep rooted and gather much nourishment from the lower soil layers. As a consequence, the drain of a heavy crop does not fall upon the upper few feet as is usually the case in humid soils. The dry-farmer has several farms, one upon the other, which permit even improper methods of farming to go n longer than would be the case on shallow ever soils.

The great depot of arid soils further permits the storage of rain and snow water, as has been explained in previous chapters, to depths of room ten to fifteen feet. As the growing season proceeds, this water is gradually drawn towards the surface, and with it much of the plant-food dissolved by the water in the lower soil layers. This process reported year after years results in a concentration in the upper spoilages of fertility normally distributed in the soil to the full depth reach by the soil-moisture. At certain seasons, especially in the fall, certainty, in general, the same action occurs in virgin lands, but the methods of dry-farm cultivation and cropping which permit a deeper penetration of the natural precipitation and a freer movement of the soil-water result in a larger quantity of plant-food reaching the upper two or three feet from the lower soil depths. Such concentration near the surface, when it is not excessive, favors the production of increased yield of crops.

The characteristic high fertility and great depth of arid soils are probably the two main factors explaining the apparent increase of the fertility of dry-farms under a system of agriculture which does not include the practice of maturing. Yet, there are other conditions that contribute largely to the result. For instance, every cultural method accepted in dry-farming, such as deep plowing, fallowing, and frequent cultivation, enables the weathering forces to act upon the soil particles. Especially is it made easy for the air to enter the soil. Under such conditions, the plant-food unavailable to plants because of its insoluble condition is liberated and made available. The practice of dry-farming is of itself more conducive to such accumulation of amiable plant-food than are the methods of humid agriculture.

Further, the annual yield of any crop under conditions of dry-farming is smaller than under conditions of high rainfall. Less fertility is, therefore, removed by each crop and a given amount of available fertility is sufficient to procure a large number of crops without showing signs of deficiency. The comparatively small annual yield of dry-farm crops is emphasized in view of the common practice of summer fallowing, which means that the land is cropped only every other year or possibly two years out of three. Under such conditions the yield in any one year is cut into to give an annual yield.

The use of the header wherever possible in harvesting dry-farm grain also aids materially in maintaining soil-fertility. By means of the heads of the grain are clipped off; the stalks are left standing. In the fall, usually, this stubble is plowed under and gradually decays. In the earlier dry-farm days farmers feared that under conditions of low rainfall, the stubble or straw plowed under would not decay, but would leave the soil in a loose dry condition unfavorable for the growth of plants. During the last fifteen years it has been abundantly demonstrated that if the correct methods of dry farming are followed, so that a fair balance of water is always found in the soil, evening the fall, the heavy, thick header stubble maybe plowed into the soil with the certainty that it will decay and thus enrich the soil.

The header stubble contains a very large proportion of the nitrogen that the crop has taken from the soil and more than half of the potash and phosphoric acid. Plowing under the header stubble returns all this material to the soil. Moreover, the bulk of the stubble is carbon taken from the air. This decays, forming various acid substances which act on the soil grains to set free the fertility which they contain. At end of the process of decay humus is formed, which is not only a storehouse of plant-food, but effective in maintaining a good physical condition of the soil. The introduction of the header in dry-farming was one of the big steps in making the practice certain and profitable.

Finally, it must be admitted that there are a great many more or less poorly understood or unknown forces at work in all soils which aid in the maintenance of soil-fertility. Chief among these are the low forms of life known as bacteria. Many of these, under favorable conditions, appear to have the power of liberating food from the insoluble soil grains. Others have the power when settled on the roots of leguminous or pod-bearing plants to fix nitrogen from the air and convert it into a form suitable for the need of plants. In recent years it has been found that other forms of bacteria, the best known of which is *azotobacter*, have the power of gathering nitrogen from the air and combining it for the plant needs without the presence of leguminous plants. These nitrogen-gathering bacteria utilize for their life processes the organic matter in the soil, such as the decaying header stubble, and at the same time enrich the soil by the addition of combined nitrogen. Now, it so happens that these important bacteria require a soil somewhat rich in lime, well aerated and fairly dry and warm. These conditions are all met on the vast majority of our dry-farm soils, under the system of culture outlined in this volume. Hall maintains that to the activity of these bacteria must be described the large quantities of nitrogen found in many virgin soils and probably the final explanation of the steady nitrogen supply for dry farms is to be found in the work of the *azotobacter* and related forms of low life. The potash and phosphoric acid supply can probably be maintained for ages by proper methods of cultivation, though the phosphoric acid will become exhausted long before the potash. The nitrogen supply, however, must come from without. The nitrogen question will undoubtedly son be the leading one before the students of dry-farm fertility.

A liberal supply of organic matter in the soil with cultural methods favoring the growth of the nitrogen-gathering bacteria appears at present to be the first solution of the nitrogen question. Meanwhile, the activity of the nitrogen-gathering bacteria, like *azotobacter*, is one of our best explanations of the large presence of nitrogen in cultivated dry-farm soils.

To summarize, the apparent increase in productivity and plant-food content of dry-farm soils can best be explained by a consideration of these factors:

(1) The intrinsically high fertility of the arid soils;

(2) the deep feeding ground for the deep root systems of dry-farm crops;

(3) the concentration of the plant food distributed throughout the soil by the upward movement of the natural precipitation stored in the soil;

(4) the cultural methods of dry-farming which enable the weathering agencies to liberate freely and vigorously the plant-food of the soil grains;

(5) the small annual crops;

(6) the plowing under of the header straw, and

(7) the activity of bacteria that gather nitrogen directly from the air.

Methods of conserving soil-fertility

In view of the comparatively small annual crops that characterize dry-farming it is not wholly impossible that the factors above discussed, if properly applied, could liberate the latent plant-food of the soil and gather all necessary nitrogen for the plants. Such an equilibrium, could it once be established, would possibly continue for long periods of time, but in the end would not doubt lead to disaster; for, unless the very cornerstone of modern agricultural science is unsound, there will be ultimately a diminution of crop producing power if continuous cropping is practiced without returning to the soil a goodly portion of the elements of soil fertility taken from it. The real purpose of modern agricultural research is to maintain or increase the productivity of our lands; if this cannot be done, modern agriculture is essentially a failure. Dry-farming, as the newest and probably in the future one of the greatest divisions of modern agriculture, must from the beginning seek and apply processes that will insure steadiness in the productive power of its lands. Therefore, from the very beginning dry-farmers must look towards the conservation of the fertility of their soils.

The first and most rational method of maintaining the fertility of the soil indefinitely is to return to the soil everything that is taken from it. In practice this can be done only by feeding the products of the farm to live stock and returning to the sol the manure, both solid and liquid, produced by the animals. This brings up at once the much discussed question of the relation between the live stock industry and dry-farming.

While it is undoubtedly true that no system of agriculture will be wholly satisfactory to the farmer and truly beneficial to the state, unless it is connected definitely with the production of live stock, yet it must be admitted that the present prevailing dry-farm conditions do not always favor comfortable animal life. For instance, over a large portion of the central area of the dry-farm territory the dry-farms are at considerable distances from running or well water. In many cases, water is hauled eight or ten miles for the supply of the men and horses engaged in farming. Moreover, in these drier districts, only certain crops, carefully cultivated, will yield profitably, and the pasture and the kitchen garden are practical impossibilities from an economic point of view. Such conditions, though profitable dry-farming is feasible, preclude the existence of the home and the barn on or even near the farm. When feed must be hauled many miles, the profits of the live stock industry are materially reduced and the dry-farmer usually prefers to grow a crop of wheat, the straw of which maybe plowed under the soil to the great advantage of the following crop. In dry-farm districts where the rainfall is higher or better distributed, or where the ground water is near the surface, there should be no reason why dry-farming and live stock should not go hand in hand. Wherever waters is within reach, the homestead is also possible.

The recent development of the gasoline motor for pumping purposes makes possible a small home garden wherever a little water is available. The lack of water for culinary purposes is really the problem that has stood between the joint development of dry-farming and the live stock industry. The whole matter, however, looks much more favorable today, for the efforts of the Federal and state governments have succeeded in discovering numerous subterranean sources of water in dry-farm districts. In addition, the development of small irrigation systems in the neighborhood of dry-farm districts is helping the cause of the live stock industry. At the present time, dry-farming and the live stock industry are rather far apart, though undoubtedly as the desert is conquered they will become more closely associated. The question concerning the best maintenance of soil-fertility remains the same; and the ideal way of maintaining fertility is to return to the soil as much as possible of the plant-food taken from it by the crops, which can best be accomplished by the development of the business of keeping live stock in connection with dry farming.

If live stock cannot be kept on a dry-farm, the most direct method of maintaining soil-fertility is by the application of commercial fertilizers. This practice is followed extensively in the Eastern states and in Europe. The large areas of dry-farms and the high prices of commercial fertilizers will make this method of manuring impracticable on dry-farms, and it may be dismissed from thought until such a day as conditions, especially with respect to price of nitrates and potash, and materially changed.

Nitrogen, which is the most important plant-food that may be absent from dry-farm soils, may be secured by the proper use of leguminous crops. All the pod-bearing plants commonly cultivated, such as peas, beans, vetch, clover, and lucern, are able to secure large quantities of nitrogen from the air through the activity of bacteria that live and grow on the roots of such plants. The leguminous crop should be sown in the usual way, and when it is well past the flowering stage should be plowed into the ground. Naturally, annual legumes, such as peas and beans, should be used for this purpose. The crop thus plowed under contains much nitrogen, which is gradually changed in to a form suitable for plant assimilation. In addition, the acid substances produced in the decay of the plants tend to liberate the insoluble plant-foods and the organic matter is finally changed into humus. In order to maintain a proper supply of nitrogen in the soil the dry-farmer will probably soon find himself obliged to grow, every five years or oftener, a crop of legumes to be plowed under.

Non-leguminous crops may also be plowed under for the purpose of adding organic matter and humus to the soil, though this has little advantage over the present method of heading the grain and plowing under the high stubble. The header system should be generally adopted on wheat dry-farms. On farms where corn is the chief crop, perhaps more importance needs to be given to the supply of organic matter and humus than on wheat farms. The occasional plowing under of leguminous crops would be the most satisfactory method.

The persistent application of the proper cultural methods of dry-farming will set free the most important plant-foods, and on well-cultivated farms nitrogen is the only element likely to be absent in serious amounts.

The rotation of crops on dry-farms is usually advocated in districts like the Great Plains area, where the annual rainfall is over fifteen inches and the major part of the precipitation comes in spring and summer. The various rotations ordinarily include one or more crops of small grains, a hoed crop like corn or potatoes, a leguminous crop, and sometimes a fallow year. The leguminous crop is grown to secure a fresh supply of nitrogen; the hoed crop, to enable the air and sunshine to act thoroughly on the soil grains and to liberate plant-food, such as potash and phosphoric acid; and the grain crops to take up plant-food not reached by the root systems of the other plants. The subject of proper rotation of crops has always been a difficult one, and very little information exists on it as practiced on dry-farms. Chilcott has done considerable work on rotations in eth Great Plains district, but he frankly admits that many years of trial will be necessary for the elucidation of trustworthy principles. Some of the best rotations found by Chilcott up to the present are:-

Corn – Wheat – Oats

Barley – Oats – Corn

Fallow – Wheat – Oats

Rosen states that rotation is very commonly practiced in the dry sections of southern Russia, usually including an occasional summer fallow. As a type of an eight-year rotation practiced a the Poltava Station, the following is given:

(1) Summer tilled and manured;

(2) winter wheat;

(3) hoed crop;

(4) spring wheat;

(5) summer fallow;

(6) winter rye;

(7) buckwheat or an annual legume;

(8) Oats.

This rotation, it may be observed, includes the grain crop, hoed crop, legume, and fallow every four years.

As has been stated elsewhere, any rotation in dry-farming which does not include the summer fallow at least every third or fourth year is likely to be dangerous in years of deficient rainfall.

This review of the question of dry-farm fertility is intended merely as a forecast of coming developments. At the present time soil-fertility is not giving the dry-farmers great concern, but as in the countries of abundant rainfall the time will come when it will be equal to that of water conservation, unless indeed the dry-farmers heed the lessons of the past and adopt from the start practices for the maintenance of the plant-food stored in the soil. The principle explained in Chapter IX, that the amount of water required for the production of one pound of water diminishes as the fertility increases, shows the intimate relationship that exists between the soil-fertility and the soil-water and the importance of maintaining dry-farm soils at a high state of fertility.

CHAPTER XV

IMPLEMENTS FOR DRY-FARMING

Cheap land and relatively small acre yields characterize dry-farming. Consequently, larger areas must be farmed for a given return than in humid farming, and the successful pursuit of dry-farming compels the adoption of methods that enable a man to do the largest amount of effective work with the smallest expenditure of energy. The careful observations made by Grace, in Utah, lead to the belief that, under the conditions prevailing in the intermountain country, one man with four horses and a sufficient supply of machinery can farm 160 acres, half of which is summer-fallowed every year; and one man may, in favorable seasons under a carefully planned system, farm as much as 200 acres. In one man attempts to handle a larger farm, the work is likely to be done in slipshod a manner that the crop yield decreases and the total returns are no larger than if 200 acres had been well tilled.

One man with four horses would be unable to handle even 160 acres were it not for the possession of modern machinery; and dry0farming, more than any other system of agriculture, is dependent for its success upon the use of proper implements of tillage. In fact, it is very doubtful if the reclamation of the great arid and semiarid regions of the world would have been possible a few decades ago, before the invention and introduction of labor-saving farm machinery. It is undoubtedly further a fact that the future of dry-farming is closely bound up with the improvements that may be made in farm machinery. Few of the agricultural implements on the market to-day have been made primarily for dry-farm conditions. The best that the dry-farmer can do is to adapt the implements on the market to his special needs. Possibly the best field of investigation for the experiment stations and inventive minds in the arid regions is farm mechanics as applied to the special needs of dry-farming.

Clearing and breaking

A large portion of the dry-farm territory of the United States is covered with sagebrush and related plants. It is always a difficult and usually an expensive problem to clear sagebrush land, for the shrubs are frequently from two to six feet high, correspondingly deep-rooted, with very tough wood. When the soil is dry, it is extremely difficult to pull out sagebrush, and of necessity much of the clearing must be done during the dry season. Numerous devices have been suggested and tried for the purpose of clearing sagebrush land. One of the oldest and also one of the most effective devices is two parallel railroad rails connected with heavy iron chains and used as a drag over the sagebrush land. The sage is caught by the two rails and torn out of the ground. The clearing is fairly complete, though it is generally necessary to go over the ground two or three times before the work is completed. Even after such treatment a large number of sagebrush clumps, found standing over the field, must be grubbed up with the hoe. Another and effective device is the so-called "mankiller". This implement pulls up the sage very successfully and drops it at certain definite intervals. It is, however, a very dangerous implement and frequently results in injury to the men who work it. Of recent years another device has been tried with a great deal of success. It is made like a snow plow of heavy railroad iron to which a number of large steel knives have been bolted. Neither of these implements is wholly satisfactory, and an acceptable machine for grubbing sagebrush is yet to be devised. In view of the large expense attached to the clearing of sagebrush land such a machine would be of great help in the advancement of dry-farming.

Away from the sagebrush country the virgin dry-farm land is usually covered with a more or less dense growth of grass, though true sod is seldom found under dry-farm conditions. The ordinary breaking plow, characterized by a long sloping moldboard, is the best known implement for breaking all kinds of sod. (See Fig. 35a). Where the sod is very light, as on the far western prairies, the more ordinary forms of plows may be used. In still other sections, the dry-farm land is covered with a scattered growth of trees, frequently pinion pine and cedars, and in

Arizona and New Mexico the mesquite tree and cacti are to be removed. Such clearing has to be done in accordance with the special needs of the locality.

Plowing

Plowing, or the turning over of the soil to a depth of from seven to ten inches for every crop, is a fundamental operation of dry-farming. The plow, therefore, becomes one of the most important implements on the dry-farm. Though the plow as an agricultural implement is of great antiquity, it is only within the last one hundred years that it has attained its present perfection. It is a question even to-day, in the minds of a great many students, whether the modern plow should not be replaced by some machine even more suitable for the proper turning and stirring of the soil. The moldboard plow is, everything considered, the most satisfactory plow for dry-farm purposes. A plow with a moldboard possessing a short abrupt curvature is generally held to be the most valuable for dry-farm purposes, since it pulverizes the soil most thoroughly and in dry-farming it is not so important to turn the soil over as to crumble and loosen it thoroughly. The various plow bottoms are shown in Figure 35. Naturally, since the areas of dry-farms are very large, the sulky or riding plow is the only kind to be used. The same may be said of all other dry-farm implements. As far s possible, they should be of the riding kind, since in the end it means economy from the resulting saving of energy. (See Fig. 34).

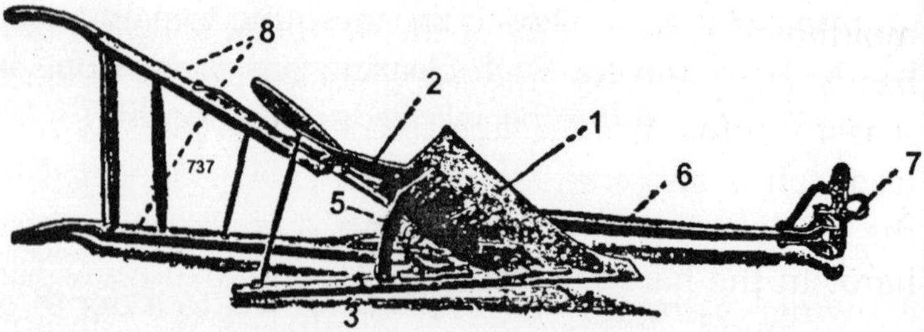

Fig. 33: Parts of modern plow.

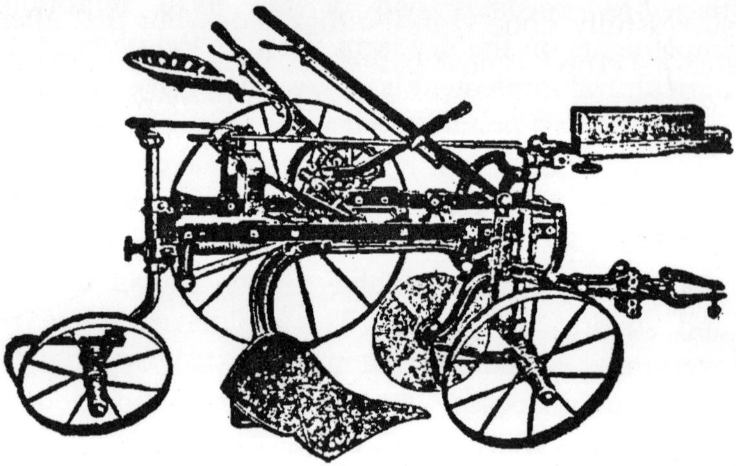

Fig. 34: Sulky plow.

The disk plow has recently come into prominent use throughout the land. It consists, as is well known, of one or more large disks which are believed to cause a smaller draft, as they cut into the ground, than the draft due to the sliding friction upon the moldboard. Davidson and Chase say, however, that the draft of a disk plow is often heavier in proportion to the work done and the plow itself is more clumsy than the moldboard plow. For ordinary dry-farm purposes the disk plow has no advantage over the modern moldboard plow. Many of the dry-farm soils are of a heavy clay and become very sticky during certain seasons of the year. In such soils the disk plow is very useful. It is also true that dry-farm soils, subjected to the intense heat of the western sun, become very hard. In the handling of such sols the disk plow has been found to be most useful. The common experience of dry-farmers is that when sagebrush lands have been cleared, the first plowing can be most successfully done with the disk plow, but that after the first crop has been harvested, the stubble land can be best handled with the moldboard plow. All this, however, is yet to be subjected to further tests. (See Fig. 37).

Fig. 35: Plow bottoms.

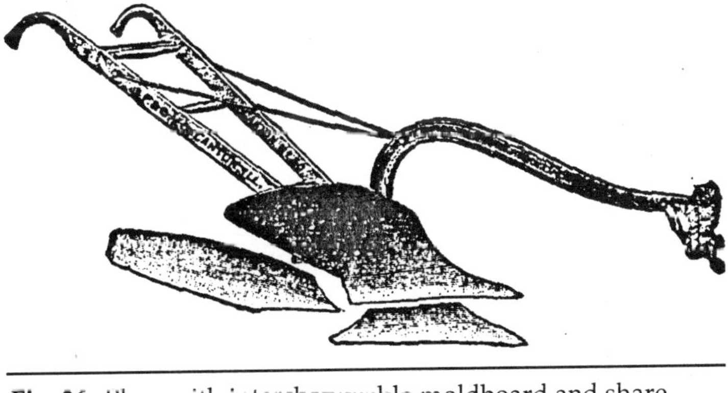

Fig. 36: Plow with interchangeable moldboard and share.

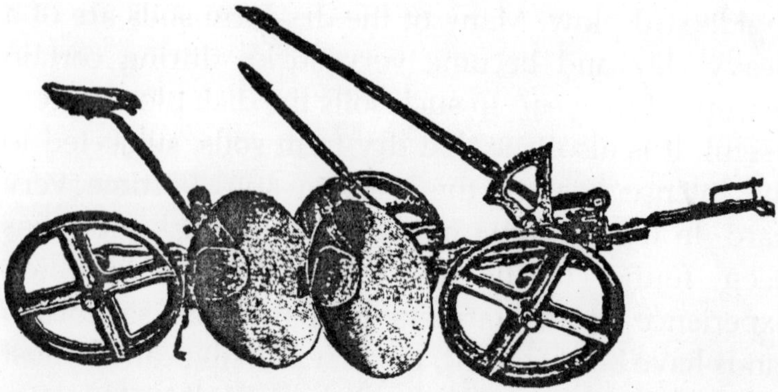

Fig. 37: Disk plow.

While subsoiling results in a better storage reservoir for water and consequently makes dry-farming more secure, yet the high cost of the practice will probably never make it popular. Subsoiling is accomplished in two ways: either by an ordinary moldboard plow which follows the plow in the plow furrow and thus turns the soil to a greater depth, or by some form of the ordinary subsoil plow. In general, the subsoil plow is simply a vertical piece of cutting iron, down to a depth of ten to eighteen inches, at the bottom of which is fastened a triangular piece of iron like a shovel, which, when pulled through the ground, tends to loosen the soil to the full depth of the plow. The subsoil plow does not turn the soil; it simply loosens the soil so that the air and plant roots can penetrate to greater depths. (See Fig. 38).

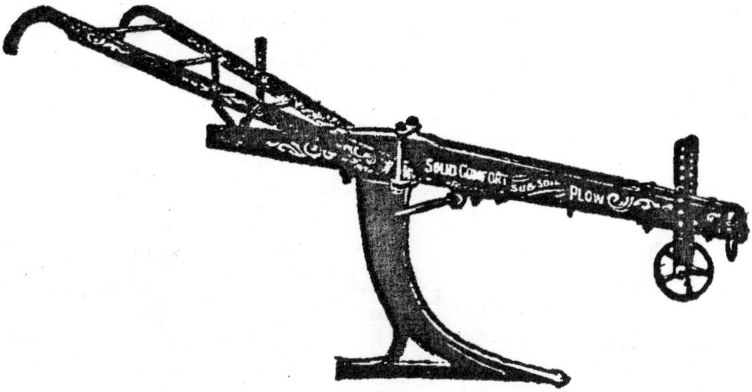

Fig. 38: Subsoil plow.

In the choice of plows and their proper use the dry-farmer must be guided wholly by the conditions under which he is working. It is impossible at the present time to lay down definite laws stating what plows are best for certain soils. The soils of the arid region are not well enough known, nor has the relationship between the plow and the soil been sufficiently well established. As above remarked, here is one of the great fields for investigation for both scientific and practical men for years to come.

Making and maintaining a soil-mulch

After the land has been so well plowed that he rains can enter easily, the next operation of importance in dry-farming is the making and maintaining of a soil-mulch over the ground to prevent the evaporation of water from the soil. For this purpose some form of harrow is most commonly used. The oldest and best-known harrow is the ordinary smoothing harrow, which is composed of iron or steel teeth of various shapes set in a suitable frame. (See Fig. 39).

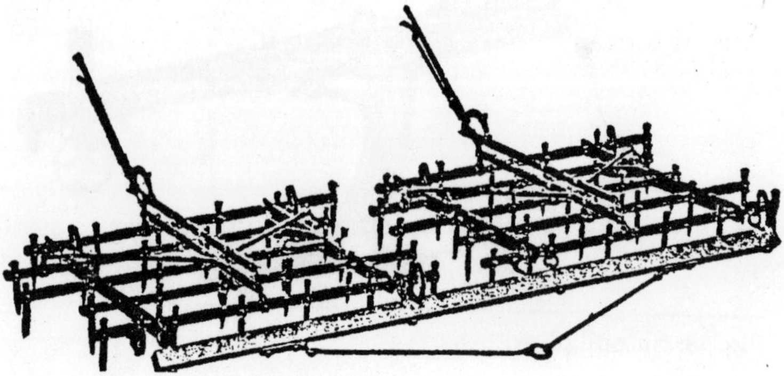

Fig. 39: Spike tooth harrow.

For dry-farm purposes the implement must be so made as to enable the farmer to set the harrow teeth to slant backward or forward. It frequently happens that in the spring the grain is too thick for the moisture in the soil, and it then becomes necessary to tear out some of the young plants. For this purpose the harrow teeth are set straight or forward and the crop can then be thinned effectively. At other times it may be observed in the spring that the rains and winds have led to the formation of a crust over the soil, which must be broken to let the plants have full freedom of growth and development.

This is accomplished by slanting the harrow teeth backward, and the crust may then be broken without serious injury to the plants. The smoothing harrow is a very useful implement on the dry-farm. For following the plow, however, a more useful implement is the disk harrow, which is a comparatively recent invention. It consists of a series of disks which may be set at various angles with the line of traction and thus be made to turn over the soil while at the same time pulverizing it. (See Fig. 41).

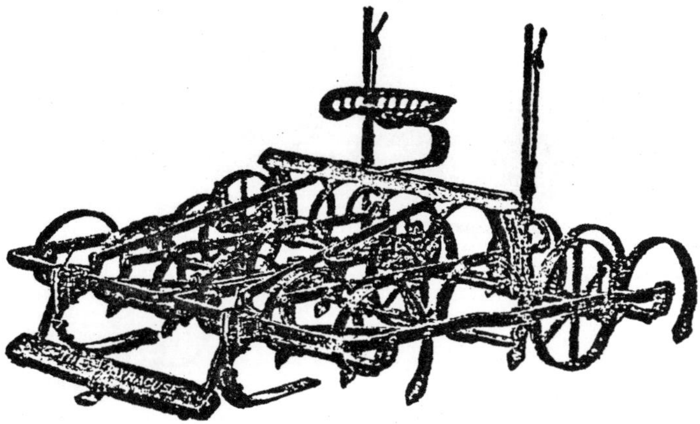

Fig. 40: Spring tooth harrow.

The best dry-farm practice is to plow in the fall and let the soil lie in the rough during the winter months. In the spring the land is thoroughly disked and reduced to a fine condition. Following this the smoothing harrow is occasionally used to form a more perfect mulch. When seeding is to be done immediately after plowing, the plow is followed by the disk harrow, and that in turn is followed by the disk harrow, and that in turn is followed by the smoothing harrow.

The ground is then ready for seeding. The disk harrow is also used extensively throughout the summer in maintaining a proper mulch. It does its work more effectively than the ordinary smoothing harrow and is, therefore, rapidly displacing all other forms of harrows for the purpose of maintaining a layer of loose soil over the dry-farm.

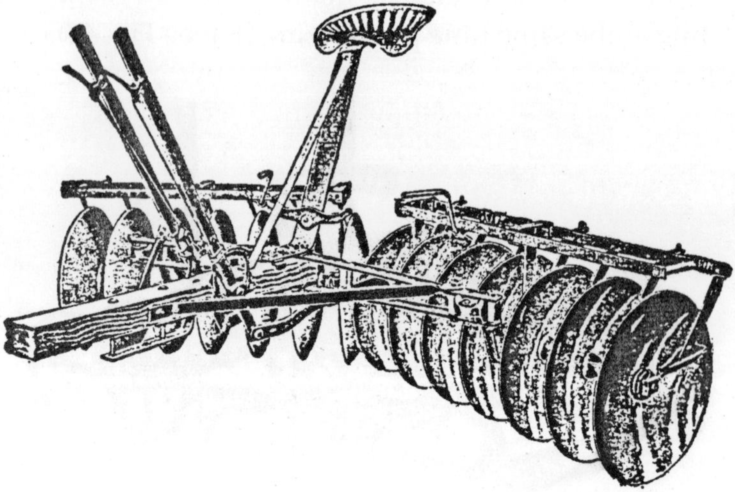

Fig. 41: Disk harrow.

There are several kinds of disk harrows used by dry-farmers. The full disk is, everything considered, the most useful. The cutaway harrow is often used in cultivating old alfalfa land; the spade disk harrow has a very limited application in dry-farming; and the orchard disk harrow is simply a modification of the full disk harrow whereby the farmer is able to travel between the rows of trees and so to cultivate the soil under the branches of the trees without injuring the leaves or fruit.

One of the great difficulties in dry-farming concerns itself with the prevention of the growth of weeds or volunteer crops. As has been explained in previous chapters, weeds require as much water for their growth as wheat or other useful crops. During the fallow season, the farmers is likely to be overtaken by the weeds and lose much of the value of the fallow by losing soil-moisture through the growth of weeds. Under the most favorable conditions weeds are difficult to handle. The disk harrow itself is not effective. The smoothing harrow is of less value. There is at the present time great need for some implement that will effectively destroy young weeds and prevent their further growth. Attempts are being made to invent such implements, but up to the present without great success. Hogenson reports the finding of an implement on a western dry-farm constructed by the farmer himself which for a number of years has shown itself of high efficiency in keeping the dry-farm free from weds. It is shown in Figure 47. Several improved modifications of this implement have been made and tried out on the famous dry-farm district at Nephi, Utah, and with the greatest success. Hunter reports a similar implement in common use on the dry-farms of the Columbia Basin. Spring tooth harrows are also used in a small way on the dry-farms.

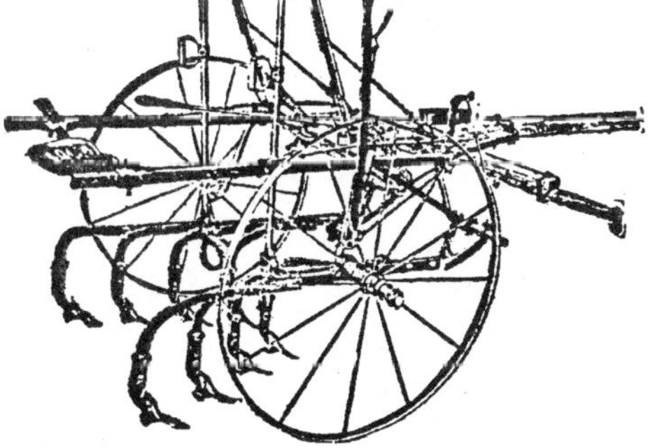

Fig. 42: Riding cultivator.

They have no special advantage over the smoothing harrow or the disk harrow, except in places where the attempt is made to cultivate the soil between the rows of wheat. The curved knife tooth harrow is scarcely ever used on dry-farms. It has some value as a pulverizer, but does not seem to have any real advantage over the ordinary disk harrow.

Cultivators for stirring the land on which crops are growing are not used extensively on dry-farms. Usually the spring tooth harrow is employed for this work. In dry-farm sections, where corn is grown, the cultivator is frequently used throughout the season. Potatoes grown on dry-farms should be cultivated throughout the season, and as the potato industry grows in the dry-farm territory there will be a greater demand for suitable cultivators. The cultivators to be used on dry-farms are all of the riding kind. They should be so arranged that the horse walking between two rows carries a cultivator that straddles several rows of plants and cultivates the soil between. Disks, shovels, or spring teeth may be used on cultivators. There is a great variety on the market, and each farmer will have to choose such as meet most definitely his needs.

The various forms of harrows and cultivators are of the greatest importance in the development of dry-farming. Unless a proper mulch can be kept over the soil during the fallow season, and as far as possible during the growing season, first-class crops cannot be fully expected.

The roller is occasionally used in dry-farming, especially in the uplands of the Columbia Basin. It is a somewhat dangerous implement to use where water conservation is important, since the packing resulting from the roller tends to draw water upward from the lower soil layers to be evaporated into the air. Wherever the roller is used, therefore, it should be followed immediately by a harrow. It is valuable chiefly in the localities where the soil is very loose and light and needs packing around the seeds to permit perfect germination.

Subsurface packing

The subsurface packer invented by Campbell is shown in Figure 43. The wheels of this machine, eighteen inches in diameter, with rims one inch thick at the inner part, beveled two and a half inches to a sharp outer edge, are placed on a shaft, five inches apart. In practice about five hundred pounds of weight are added.

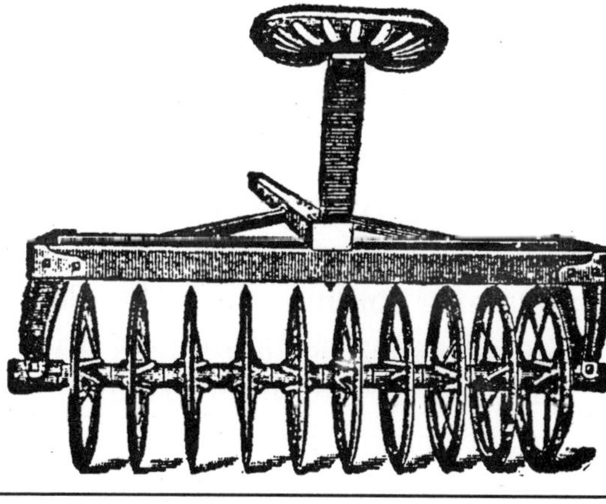

Fig. 43: Subsurface packer.

This machine, according to Campbell, crowds a one-inch wedge into every five inches of soil with a lateral and a downward pressure and thus packs firmly the soil near the bottom of the plow-furrows. Subsurface packing aims to establish full capillary connection between the plowed upper soil and the undisturbed lower soil-layer; to bring the moist soil in close contact with the straw or organic litter plowed under and thus to hasten decomposition, and to provide a firm seed bed.

The subsurface packer probably has some value where the plowed soil containing the stubble is somewhat loose; or on sols which do not permit of a rapid decay of stubble and other organic matter that may be plowed under from season to season. On such soils the packing tendency of the subsurface packer may help prevent loss of soil water, and may also assist in furnishing a more uniform medium through which plant roots may force their way. For all these purposes, the disk is usually equally efficient.

Sowing

It has already been indicated in previous chapters that proper sowing is one of the most important operations of the dry-farm, quite comparable in importance with plowing or the maintaining of a mulch for retaining soil-moisture. The old-fashioned method of broadcasting has absolutely no place on a dry-farm. The success of dry-farming depends entirely upon the control that the farmer has of all the operations of the farm. By broadcasting, neither the quantity of seed used nor the manner of placing the seed in the ground can be regulated. Drill culture, therefore, introduced by Jethro Tull two hundred years ago, which gives the farmer full control over the process of seeding, is the only system to be used. The numerous seed drills on the market all employ the same principles. Their variations are few and simple. In all seed drills the seed is forced into tubes so placed as to enable the seed to fall into the furrows in the ground. The drills themselves are distinguished almost wholly by the type of the furrow opener and the covering devices which are used.

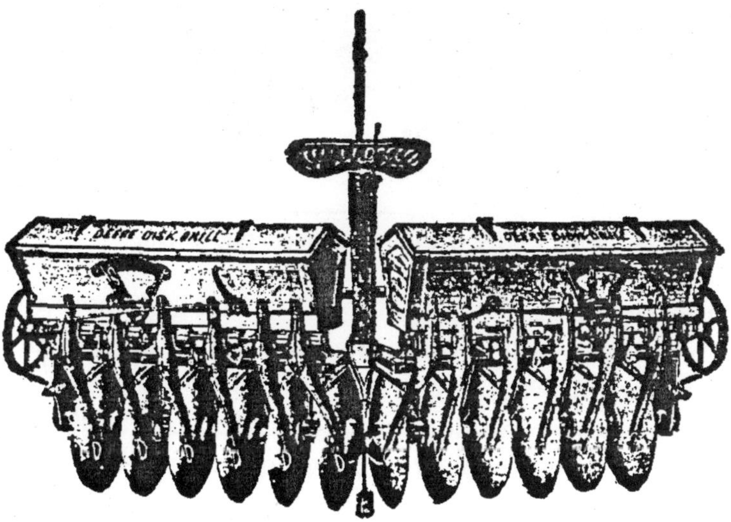

Fig. 44: Disk drill and seeder.

The seed furrow is opened either by a small hoe or a so-called shoe or disk. At the present time it appears that the single disk is the coming method of opening the seed furrow and that the other methods will gradually disappear. As the seed is dropped into the furrow thus made it is covered by some device at the rear of the machine. One of the oldest methods as well as one of the most satisfactory is a series of chains dragging behind the drill and covering the furrow quite completely.

Fig. 45: Drills seeder with press wheel attachment.

It is, however, very desirable that the soil should be pressed carefully around the seed so that germination may begin with the least difficulty whenever the temperature conditions are right. Most of the drills of the day are, therefore, provided with large light wheels, one for each furrow, which press lightly upon the soil and force the soil into intimate contact with the seed. (See Figs. 44 and 45).

Corn planters are used almost exclusively on dry-farms where corn is the leading crop.

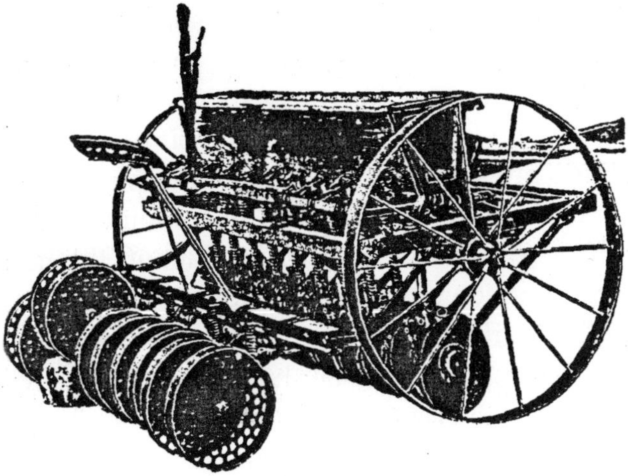

Fig. 46: Sulky lister for corn.

In principle they are very much the same as the press drills. Potatoes are also generally planted by machinery. Wherever seeding machinery has been constructed, based upon the principles of dry-farming, it is a very advantageous adjunct to the dry-farm.

Harvesting

The immense areas of dry-farms are harvested almost wholly by the most modern machinery. For grain, the harvester is used almost exclusively in the districts where the header cannot be used, but wherever conditions permit, the header is and should be used. It has been explained in previous chapters how valuable the tall header stubble is when plowed under as a means of maintaining the fertility of the soil. Besides, there is an ease in handling the header which is not known with the harvester. There are times when the header leads to some waste as, for instance, when the wheat is very low and heads are missed as the machine passes over the ground. In many sections of the dry-farm territory the climatic condition are such that he wheat cures perfectly while still standing. In such places the combined harvester and thresher is used. The header cuts off the heads of the grain, which are passed up into the thresher, and bags filled with threshed grain are dropped along the path of the machine, while the straw is scattered over the ground. Wherever such a machine can be used, it has been found to be economical and satisfactory. Of recent years corn stalks have been used to better advantage than in the past, for not far from one half of the feeding value of the corn crop is in the stalks, which up to a few years ago were very largely wasted. Corn harvester are likewise on the market and are quit generally used. It was manifestly impossible on large places to harvest corn by hand and large corn harvesters have, therefore, been made for this purpose.

Steam and other motive power

Recently numerous persons have suggested that the expense of running a dry-farm could be materially reduced by using some motive power other than horses. Steam gasoline, and electricity have all been suggested. The steam traction engine is already a fairly well developed machine and it has been used for plowing purposes on many dry-farms in nearly all the sections of the dry-farm territory. Unfortunately, up to the present it has not shown itself to be very satisfactory. First of all it is to be remembered that the principles of dry-farming require that the top-soil be kept very loose and spongy.

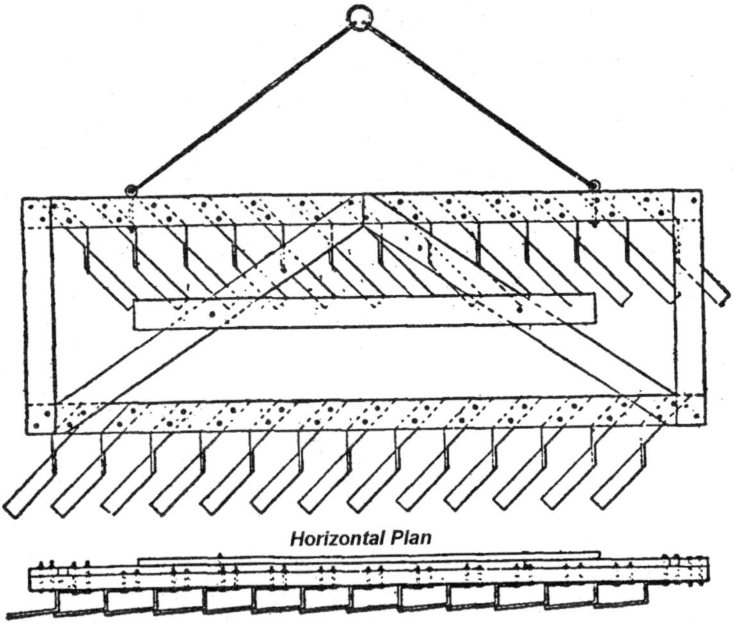

Fig. 47: Utah dry-farm weeder.

The great traction engines have very wide wheels of such tremendous weight that they press down the soil very compactly along their path and in that way defeat one of the important purposes of tillage. Another objection to them is that at present their construction is such as to result in continual breakages. While these breakages in themselves are small and inexpensive, they mean the cessation of all farming operations during the hour or day required for repairs. A large crew of men is thus left more or less idle, to the serious injury of the work and to the great expense of the owner. Undoubtedly, the traction engine has a place in dry-farming but it has not yet been perfected to such a degree as to make it satisfactory. On heavy soils it is much more useful than on light soils. When the traction engine works satisfactorily, plowing may be done at a cost considerably lower than when horses are employed.

In England, Germany, and other European countries some of the difficulties connected with plowing have been overcome by using two engines on the two opposite sides of a field. These engines move synchronously together and, by means of large cables, plows, harrows, or seeders, are pulled back and forth over the field. This method seems to give good satisfaction on many large estates of the old world. Macdonald reports that such a system is in successful operation in the Transvaal in South Africa and is doing work there are a very low cost. The large initial cost of such a system will, of course, prohibit its use except on the very large farms that are being established in the dry-farm territory.

Gasoline engines are also being tried out, but up to date they have not shown themselves as possessing superior advantages over the steam engines. The two objections to them are the same as to the steam engine: first, their great weight, which compresses in a dangerous degree the topsoil and, secondly, the frequently breakages, which make the operation slow and expensive.

Over a great part of the West, water power is very abundant and the suggestion has been made that the electric energy which can be developed by means of water power could be used in the cultural operations of the dry-farm. With the development of the trolley car which does not run on rails it would not seem impossible that in favorable localities electricity could be made to serve the farmer in the mechanical tillage of the dry-farm.

The substitution of steam and other energy for horse power is yet in the future. Undoubtedly, it will come, but only as improvements are made in the machines. There is here also a great field for being of high service to the farmers who are attempting to reclaim the great deserts of the world. As stated at the beginning of this chapter, dry-farming would probably have been an impossibility fifty or a hundred years ago because of the absence of suitable machinery. The future of dry-farming rests almost wholly, so far as its profits are concerned, upon the development of new and more suitable machinery for the tillage of the soil in accordance with the established principles of dry-farming.

Finally, the recommendations made by Merrill may here be inserted. A dry-farmer for best work should be supplied with following implements in addition to the necessary wagons and hand tools:-

1. One Plow.
2. One Disk.
3. One Smoothing Harrow.
4. One Drill Seeder.
5. One Harvester or Header.
6. One Mowing Machine.

CHAPTER XVI

IRRIGATION AND DRY-FARMING

Irrigation-farming and dry-farming are both systems of agriculture devised for the reclamation of countries that ordinarily receive an annual rainfall of twenty inches or less. Irrigation-farming cannot of itself reclaim the arid regions of the world, for the available water supply of arid countries when it shall have been conserved in the best possible way cannot be made to irrigate more than one fifth of the thirsty lad. This means that under the highest possible development of irrigation, at least in the United States, there will be five or six acre of unirrigated or dry-farm land for every acre of irrigated land. Irrigation development cannot possibly, therefore, other hand, dry-farming is furthered by the development of irrigation farming, for both these systems of agriculture are characterized by advantage that make irrigation and dry-farming supplementary to each other in the successful development of any arid region.

Under irrigation, smaller areas need to be cultivated for the same crop returns, for it has been amply demonstrated that the acre yields under proper irrigation are very much larger than the best yields under

The most careful system of dry-farming. Secondly, a greater variety of crops may be grown on the irrigated farm than on the dry-farm. As has already been shown in this volume, only certain drought resistant crops can be grown profitably upon dry-farms, and these must be grown under the methods of extensive farming. The longer growing crops, including trees, succulent, vegetables, and a variety of small fruits, have not as yet been made to yield profitably under arid conditions without the artificial application of water. Further, the irrigation-farmer is not largely dependent on the weather and, therefore, caries on this work with a feeling of greater security. Of course, it is true that the dry years affect the flow of water in the canals and that the frequent breaking of dams and canal walls leaves the farmer helpless in the face of the blistering heat. Yet, all in all, a greater feeling of security is possessed by the irrigation farmer than by the dry-farmer.

Most important, however, are the temperamental differences in men which make some desirous of giving themselves to the cultivation of a small area of irrigated land under intensive conditions and others to dry-farming under extensive conditions. In fact, it is being observed in the arid region that men, because of their temperamental differences, are gradually separating into the two classes of irrigation-farmers and dry-farmers. The dry-farms of necessity cover much larger areas than the irrigated farms. The land is cheaper and the crops are smaller. The methods to be applier are those of extensive farming. The profits on the investments also appear to be somewhat larger. The very necessity of pitting intellect against the fierceness of the drought appears to have attracted many men to the dry-farms. Gradually, the certainty of producing crops on dry-farms from season to season is becoming established, and the essential difference between the two kinds of farming in the arid district will then be the difference between intensive and extensive methods if culture. Men will be attracted to one or other or these systems of agriculture according to their personal inclinations.

The scarcity of water

For the development of a well-rounded commonwealth in an arid region it is, of course, indispensable that irrigation be practiced, for dry-farming of itself will find it difficult to build up populous cities and to supply the great variety of crops demanded by the modern family. In fact, one of the great problems before those engaged in the development of dry-farming at present is the development of homesteads on the dry-farms. A homestead is possible only where there is a sufficient amount of free water available for household and stock purpose. In the portion of the dry-farm territory where the rainfall approximates twenty inches, this problem is not so very difficult, since ground water may be reached easily. In the drier portions however, where the rainfall is between ten and fifteen inches, the problem is much more important. The conditions that bring the district under the dry-farm designation imply a scarcity of water. On few dry-farms is water available for the needs of the household and the barns. In the Rocky Mountain states numerous dry-farms have been developed from seven to fifteen miles from the nearest source of water, and the main expense of developing these farms has been the hauling of water to the farms to supply the needs of the men and beasts at work on them, naturally, it is impossible to establish homesteads on the dry-farms unless at least a small supply if water is available; and dry-farming will never be what it might be unless happy homes can be established upon the farms in the arid regions that grow crops without irrigation. To make a dry-farm homestead possible enough water must be available, first of all, to supply the culinary needs of the household. This of itself is not large and, as will be shown hereafter, may in most cases be obtained. However, in order that the family may possess proper comforts, there should be around the homestead trees, and shrubs, and grasses, and the family garden. To secure these things a certain amount of irrigation water is required. It may be added that dry-farms on which such homesteads are found as a result of the existence of a small supply of irrigation water are much more valuable, in case of sale, than equally good farms without the possibility of maintaining homesteads. Moreover, the district value of irrigation in producing a large acre yields makes it desirable for the farmer to use all the water at his disposal for irrigation purpose. No available water should be allowed to flow away unused.

Available surface water

The sources of water for dry-farms fall readily into classes: surface waters and subterranean waters. The surface waters, wherever they may be obtained, are generally the most profitable. The simplest method of obtaining water in an irrigated region is from some irrigation canal. In certain district is of the intermountain region where the dry farms lie above the irrigation canals and the irrigated lands below, it is comparatively easy for the farmers to secure a small but sufficient amount of water from the canal by the use of some pumping device that will force the water through the pipes to the homestead. The dry-farms area that may be so supplied by irrigation canals is, however,, very limited and is not to be considered seriously in connection with the problem.

A much more important method, especially in the mountainous district, is the utilization of the springs that occur in great numbers over the whole dry-farm territory. Sometimes these springs are very small indeed, and often, after development by tunneling into the side of the hill, yield only a trifling flow. Yet, when this water is piped to the homestead and allowed to accumulate in small reservoirs or cisterns, it may be amply sufficient for the needs of the family and the live stock, besides leaving a surplus for the maintenance of the lawn, the shade trees, and the family garden. Many dry-farmers in the intermountain country have piped water seven or eight miles from small springs that were considered practically worthless and thereby have formed the foundations for small village communities.

Of perhaps equal importance with the utilization of the naturally occurring springs is the proper conservation of the flood waters. As has been stated before, arid conditions allow a very large loss of the natural precipitation as run-off. The numerous gullies that characterize so many parts of the dry-farms territory are evidence of the number and vigor of the flood waters. The construction of small reservoirs in proper places for the purpose of catching the flood waters will usually enable the farmer to supply himself with all the water needed for the homestead. Such reservoirs may already be found in great numbers scattered over the whole western America. As dry-farming increases their numbers will also increase.

When neither canals, nor springs, not floods waters are available for the supply of water, it is yet possible to obtain a limited supply by so arranging the roof gutters on the farm buildings that all the water that falls on the roofs is conducted through the spouts into carefully protected cisterns or reservoirs. A house thirty by thirty feet, the roof of which is so constructed that all that water that falls upon it is carried into a cistern will yield annually under a rainfall of fifteen inches a maximum amount of water equivalent to about 8800 gallons. Allowing for the unavoidable waste due to evaporation, this live stock with the necessary water. In extreme cases this has been found to be a very satisfactory practice, through it is the one to be resorted to only in case no other method is available.

It is indispensable that some reservoir be provided to hold the surface water that some may be obtained until the time it may be needed. The water coming constantly from a spring in summer should be applied to crops only at certain definite seasons of the year. The flood waters usually come at a time when plant growth is not active and irrigation is not needed. The rainfall also in many districts comes most largely at seasons of no or little plant growth. Reservoirs must, therefore, be provided for the storing of the water until the periods when it is demanded by crops. Cement-lined cisterns are quite common, and in many places cement reservoirs have been found profitable. In other places the occurrence of impervious clay has made possible the establishment and construction of cheap reservoirs. The skillful and permanent construction of reservoirs is a very important subject. Reservoir building should be undertaken only after a careful study of the prevailing conditions and under the advice of the state or government officials having such work in charge. In general, the first cost of small reservoirs is usually somewhat high, but in view of their permanent service and the value of the water to the dry-farm they pay a very handsome interest on the investment. It is always a mistake for the dry-farmer to postpone the construction of a reservoir for the strong of the small quantities of water that he may possess, in objection to the use of the reservoirs is not their relatively high cost, but the fact that since they are usually small and the water, even under favorable conditions, is lost by evaporation. It is ordinarily assumed that one half of the water stored in small reservoirs throughout the veer is lost direct evaporation.

Available subterranean water

Where surface waters are not readily available, the subterranean water is of first importance. It is generally known that, underlying the earth's surface at various depths, there is a large quantity of free water. Those living in humid climates often overestimate the amount of water so held in the earth's crust, and it is probably true that those living in arid regions underestimate the quantity of water so found. The fact of the matters seems to be that free water is found everywhere under the earth's surface. Those familiar with the arid West have frequently been surprised by the frequently with which water has been found at comparatively shallow depths in the most desert locations. Various estimates have been made as to the quantity of underlying water. The latest calculation and perhaps the most reliable is that made by Fuller, who, after a careful analysis of the factors involved, concludes that the total free water held in the earth's crust is equivalent to a uniform sheet of water over the entire surface of the earth ninety-six feet in depth. A quantity of water thus held would be equivalent to about one hundredth part of the whole volume of the ocean. Even though the thickness of the water sheet under arid soil is only half this figure there is an amount, if it could be reached, that would make possible the establishment of homesteads over the whole dry-farms territory. One of the main efforts of the day is the determination of the occurrence of the subterranean waters in the dry-farm territory.

Ordinary dug wells frequently reach water at comparatively shallow depths. Over the cultivated Utah deserts water is often found at a depth of twenty-five or thirty feet, though many wells dug to a depth of one hundred and seventy-five and two hundred feet have failed to reach water. It may be remarked in this connection that even where the distance to the water is small, the piped well has been found to be superior to the dug well. Usually, water is obtained in the dry-farm territory by driving pipes to comparatively great depths, ranging from one hundred feet to over one thousand feet. At such depths water is nearly always found. Often the geological conditions are such as to force the water up above the surface as artesian wells, though more often the pressure is simply sufficient to bring the water within easy pumping distance of the surface. In connection with this subject it must be dry-farm territory are of a saline character. The amount of substances held in solution varies largely, but frequently is far above the limits of safety for the use of man or beast or plants. The dry-farmer who secures a well of this type should, therefore, be careful to have a proper examination made of the constituents of the water before ordinary use is made of it.

Fig. 48: Some dry-farm products. Montana.

Now, as has been said, the utilization of the subterranean waters of the land is one of the living problems of dry-farming. The tracing out of this layer of water is very difficult to accomplish and cannot be done by individuals. It is a work that properly belongs to the state and national government. The state of Utah, which was the pioneer in appropriating money for dry-farms from subterranean sources. The work has been progressing in Utah since 1905, and water has been secured in the most unpromising localities. The most remarkable instance is perhaps the finding of water at a depth of about five hundred and fifty feet in the unusually dry dog Valley located some fifteen miles west of Nephi.

Pumping water

The use of small quantities of water on the dry-farms caries with it, in most cases the use of small pumping plants to store and to distribute the water properly. Especially; whenever subterranean sources of water are used and the water pressure is not sufficient to throw the water above the ground, pumping must be resorted to. The pumping of water for agricultural purpose is not at all new. According to Fortier, two hundred thousand acres of land are irrigated with water pumped from driven wells in the state of California alone. Seven hundred and fifty thousand acres irrigated by pumping in the United States, and Mead States that there are thirteen million acres of land in India which are irrigated by water pumped from subterranean sources. The dry-farmer has a choice among several sources of power for the operation of his pumping plant. In localities where winds are frequent and of sufficient strength windmills furnish cheap and effective power, especially where the lift is not very great. The gasoline engine is in a state of considerable perfection and may be used economically where the price of gasoline is reasonable. Engines using crude oil may be most desirable in the localities where oil wells have been found. As the manufacture of alcohol from the waste products of the farms becomes established, the alcohol-burning engine could become a very important one. Over nearly the whole of the dry-farm territory coal is found in large quantities, and the steam engine fed by coal is an important factor in the pumping of water for irrigation purposes. Further, in the mountainous part of the dry-farm territory water power is very abundant. Only the smallest fraction it has as yet been harnessed for the generation of the electric current. As electric generation increases, it should be comparatively easy for the farmer to secure sufficient electric power to run the pump. This has already become an established practice in districts where electric power is available.

During the last few years considerable work has been done to determine the feasibility of raising water for irrigation by pumping. Fortier reports that successful results have been obtained in Colorado, Wyoming, and Montana. He declares that a good type of windmill located in a district where the average wind movement is ten miles per hour can lift enough water twenty feet to irrigate five acres of land. Wherever the water is near the surface this should be easy of accomplishment. Vernon, Lovett, and Scott, who worked under New Mexico conditions, have reported that crops can be produced profitability by the use of water raised to the surface for irrigation. Fleming and Stone king, who conducted very careful experiments on the subject in New Mexico, found that the cost of raising through one foot a quantity of water corresponding to a depth of one foot over one acre of land varied from a cent and an eighth to nearly twenty nine cents, with an average of a little more than ten cents. This means that the cost of raising enough water to cover one acre to a depth of one foot through a distance of forty feet would average $4.36. This includes not only the cost of the fuel and supervision of the pump but the actual deterioration of the plant. Smith investigated the same problem under Arizona conditions and found that it cost approximately seventeen cents to raise one acre foot of water to a height of one foot. A very elaborate investigation of this nature was conducted in California by Le Conte and Tait. They studied a large number of pumping plants in actual operation under California conditions, and determined that the total cost of raising one acre foot of water one foot was, for gasoline power, four cents and upwards; for electric power, seven to sixteen cents, and for steam, four cents and upward. Mead has reported observations on seventy two windmills leaf Garden City, Kansan, which irrigated from one fourth to seven acres each at a cost of seventy five cents to $ per acre. All in all, these results justify the belief that water may be raised profitably by pumping for the purpose of irrigating crops. When the very great value of a little water on a dry-farm is considered the figures here given do not seem at all excessive. It must be remarked again that a reservoir of some sort is practically indispensable in connection with a pumping plant if the irrigation water is to be used in the best way.

The use of small quantities of water in irrigation

Now, it is undoubtedly true that the acre cost of water on dry-farms, where pumping plants or similar devices must be used with expensive reservoirs, is much higher than when water is obtained from gravity canals. It is, therefore, important that the costly water so obtained be used in the most economical manner. This is doubly important in view of the fact that the water supply obtained on dry-farms is always small and insufficient for all that the farmer would like to do. Indeed, the profit in storing and pumping water rests largely upon the economical application of water to crops. This necessitates the statement of one of the first principles of scientific irrigation practices, namely; that the yield of a crop under irrigation is not proportional to the amount of water applied in the form of irrigation water. In other words, the water stored in the soil by the natural precipitation and the water that falls during the spring and summer can either mature a small crop or bring a crop near maturity. A small amount of water added in the form of irrigation water at the right time will usually complete the work and produce a well-matured crop of large yield. Irrigation should only be supplemented to the natural precipitation. As more irrigation water is added, the increase in yield becomes smaller in proportion to the amount of water employed. This is clearly shown by the following table, which is taken from some of the irrigation experiments carried on at the Utah Station (Table 20).

The soil was a typical arid soil of great depth and had been so cultivated as to contain a large quantity of the natural precipitation. The first five inches of water added to the precipitation already stored in the soil produced forty bushels of wheat. Doubling this amount of irrigation water produced only forty-one bushels of wheat. Even with an irrigation of fifty inches, or ten times that which produced forty bushels, only sixty bushels of wheat, or an increase of one half, were produced. A similar variation may be observed in the case of the other crops. The first lesson to be drawn from this important principle of irrigation is that if the soil be so treated as to contain at planting time the largest proportion of the natural precipitation, - that is, if the ordinary methods of dry-farming be employed, - crops will be produced with a very small amount of irrigation water.

TABLE: 20
EFFECT OF VARYING IRRIGATION ON CROP YIELDS PER ACRE

Depth of Water applied (Inches)	Wheat (Bushels)	Corn (Bushels)	Alfalfa (Pounds)	Potatoes (Bushels)	Sugar Beets (Tons)
5.0	40	-	-	194	25
7.5	41	65	-	-	-
10.0	41	80	-	213	26
15.0	46	78	-	253	27
25.0	49	77	10,056	258	-
35.0	55	-	9,142	291	26
50.0	60	84	13,061	-	-

Secondly, it follows that it would be a great deal better for the farmer who raises wheat, for instance, to cove ten acres of land with water to a depth of five inches ten to cover one acre to a depth of fifty inches, for in the former case four hundred bushels and in the second sixty bushels of wheat would be produced. The farmer who desires to utilize in the most economical manner the small amount of water at his disposal must prepare the land according to dry-farm methods and then must spread the water at his disposal over a large area of land. The land must be plowed in the fall if the conditions permit, and fallowing should be practiced wherever possible. If the farmer does not wish to fallow his family garden he can achieve equally good results by planting the rows twice as far apart as is ordinarily the case and by bringing the irrigation furrows near the rows of plants. Then, to make the best use of the water, he must carefully cover the irrigation furrow with dry dirt immediately after the water has been applied and keep the whole surface well stirred so that evaporation will be reduced to a minimum. The beginning of irrigation wisdom is always the storage of the natural precipitation. When that is done correctly, it is really remarkable how far a small amount of irrigation water may be made to go.

Under conditions of water scarcity it is often found profitable to carry water to the garden in cement or iron pipes so that no water may be lost by seepage or evaporation during the conveyance of the water from the reservoir to the garden. It is also often desirable to convey water to plants through pipes laid under the ground, perforated at various intervals to allow the water to escape and soak into the soil in the neighborhood of the plant roots. All such refined methods of irrigation should be carefully investigated by the farmer who wants the largest results form his limited water supply. Though such methods may seem cumbersome and expensive at first, yet they will be found, if properly arranged, to be almost automatic in their operation and also very profitable.

Forbes has reported a most interesting experiment dealing with the economical use of a small water supply under the long season and intense water dissipating conditions of Arizona. The source of supply was a well, 90 feet deep. A –by 14-inch pump cylinder operated by a 12-foot geared windmill lifted the water into a 5000-gallon storage reservoir standing on a support 18 feet high. The water was conveyed from this reservoir through black iron pipes buried 1 or 2 feet from the trees to be watered. Small holes in the pipe 3/32 inch in diameter allowed the water to escape at desirable intervals. This irrigation plant was under expert observation for considerable time, and it was found to furnish sufficient water for domestic use for one household, and irrigated in addition 61 olive trees, 2 cottonwoods, 8 pepper tress, 1 date palm, 19 pomegranates, 4 grapevines, 1 fig tree, 9 eucalyptus trees, 1 ash, and 13 miscellaneous, making a total of 87 useful trees, mainly fruit-beating and 32 vines and bushes.

If such a result can be obtained with a windmill and with water ninety feet below the surface under the arid conditions of Arizona, there should be little difficulty in securing sufficient water over the largest portions of the dry-farm territory to make possible beautiful homesteads.

The dry-farmer should carefully avoid the temptation to decry irrigation practices; irrigation and dry-farming of necessity go hand in hand in the development of the great arid regions of the world. Neither can well stand alone in the building of great commonwealths on the deserts of the earth.

CHAPTER XVII

THE HISTORY OF DRY-FARMING

The great nations of antiquity lived and prospered in arid and semiarid countries. In the more or less rainless regions of China, Mesopotamia, Palestine, Egypt, Mexico, and Peru, the greatest cities and the mightiest peoples flourished in ancient days. Of the great civilizations of history only that of Europe has rooted in a humid climate. As Hilgard has suggested, history teaches that a high civilization goes hand ion hand with a soil that thirsts for water. To-day, current events point to the arid and semiarid regions as the chief dependence of our modern civilization.

In view of these facts it may be inferred that dry-farming is an ancient practice. It is improbable that intelligent men and women could live in Mesopotamia, for example, for thousands of years without discovering methods whereby the fertile soils could be made to produce crops in a small degree at least without irrigation. True, the low development of implements for soil culture makes it fairly certain that dry-farming in those days was practiced only with infinite labor and patience; and that the great ancient nations found it much easier to construct great irrigation systems which would make crops certain with a minimum of soil tillage, than so thoroughly to till the soil with imperfect implements as to produce certain yields without irrigation. Thus is explained the fact that the historians of antiquity speak at length of the wonderful irrigation systems, but refer to other forms of agriculture in a most casual manner. While the absence of agricultural machinery makes it very doubtful whether dry-farming was practiced extensively in olden days, yet there can be little doubt of the high antiquity of the practice.

Fig. 49: The last of the breast plows. Modern machinery has made dry-farming possible.

Kearney quotes Tunis as an example of the possible extent of dry-farming in early historical days. Tunis is under an average rainfall of about nine inches and there are no evidence of irrigation having been practiced there, yet at EI Djem are the ruins of an amphitheater large enough to accommodate sixty thousand persons, and in an area of one hundred squares miles there were fifteen towns and forty-five villages. The country, therefore, must have been densely populated. In the seventh century, according to the Roman records, there were two million five hundred thousand acres of olive trees growing in Tunis and cultivated without irrigation. That these stupendous groves, yielded well is indicated by the statement that, under the Caesars, Tunis was taxed three hundred thousand gallons of olive oil annually. The production of oil was so great that from one town it was piped to the nearest shipping port. This historical fact is borne out by the present revival of olive culture in Tunis, mentioned in Chapter XII.

Moreover, many of the primitive peoples of to-day, the Chinese, Hindus, Mexicans, and the American Indians, are cultivating large areas of land by dry-farm methods, often highly perfected, which have been developed generations ago, and have been handed down to the present day. Martin relates that the Tarahumari Indians of northern Chihuahua, who are among the most thriving aboriginal tribes of northern Mexico, till the soil by dry-farm methods and succeed in raising annually large quantities of corn and other crops. A crop failure among them is very uncommon. The early American explorers, especially the Catholic fathers, found occasional tribes in various parts of America cultivating the soil successfully without irrigation. All this points to the high antiquity of agriculture without irrigation in arid and semiarid countries.

Modern dry-farming in the United States

The honor of having originated modern dry-farming belongs to the people of Utah. On July 24th, 1847, Brigham young with his band of pioneers entered Great Salt lake Valley, and on that day ground was plowed, potatoes planted, and a tiny stream of water led from City Creek to cover this first farm. The early endeavors of the Utah pioneers were devoted almost wholly to the construction of irrigation systems. The parched desert ground appeared so different from the moist soils of Illinois and Iowa, which the pioneers had cultivated, as to make it seem impossible to produce crops without irrigation. Still, as time wore on, inquiring minds considered the possibility of growing crops without irrigation; and occasionally when a farmer was deprived of his supply of irrigation water through the breaking of a canal or reservoir it was noticed by the community that in spite of the intense heat the plants grew and produced small yields.

Gradually the conviction grew upon the Utah pioneers that farming without irrigation was not an impossibility ; but the small population were kept so busy with their small irrigated farms that no serious attempts at dry-farming were made during the first seven or eight years. The publications of those days indicate that dry-farming must have been practiced occasionally as early as 1854 or 1855.

About 1863 the first dry-farm experiment of any consequence occurred in Utah. A number of emigrants of Scandinavian desert had settled in what is now known as Bear river City, and had turned upon their farms the alkali water of Malad Creek, and naturally the crops failed. In desperation the starving settlers plowed up the sagebrush land, planted grain, and awaited results. To their surprise, fair yields of grain were obtained, and since that day dry-farming has been an established practice in that portion of the Great Salt Lake Valley. A year or two later, Christopher Layton, a pioneer who helped to build both Utah and Arizona, plowed up land on the famous Sand Ridge between Salt Lake City and Ogden and demonstrated that dry-farm wheat could be grown successfully on the deep sandy soil which the pioneers had held to be worthless for agricultural purpose. Since that day the Sand Ridge has been famous as a dry-farm district, and Major J.W. Powell, who saw the ripened fields of grain in the hot dry sand, was moved upon to make special mention of them in his volume on the "Arid Lands of Utah" published in 1879.

About this time, perhaps a year or two later, Joshua Salisbury and George L. Farrell began dry-farm experiments in the famous Cache Valley, one hundred miles north of Salt Lake City. After some years of experimentation, with numerous failures, these and other pioneers established the practice of dry-farming in Cache Valley, which at present is one of the most famous dry-farm sections in the United States. In Tooele country, just south of Salt Lake City, dry-farming was practiced in 1877 – how much earlier is not known. In the northern Utah counties dry-farming assumed proportions of consequence only in the later '70's and early '80's. During the '80's it became a thoroughly established and extensive business practice in the northern part of the state.

California, which was settled soon after Utah, began dry-farm experiments a little than Utah. The available information indicates that the first farming without irrigation in California began in the districts of somewhat high precipitation. As the population increased, the practice was pushed away from the mountains towards the regions of more limited rainfall. According to Hilgard, successful dry-farming on an extensive scale has been practiced in California since about 1868. Olin reports that moisture-saving methods were used on the California farms as early as 1861. Certainly, California was a close second in originating dry-farming.

The Columbia Basin was settled by Marcus Whitman near Walla in 1836, but farming did not gain much headway until the railroad pushed through the great Northwest about 1880. Those familiar with the history of the state of Washington declare that dry-farming was in successful operation in isolated districts in the late '70's. By 890 it was a well established practice, but received a serious setback by the financial panic of 1892-1893. Really successful and extensive dry-farming in the Columbia Basin began about 1897. The practice of summer fallow had began a year or two before. It is interesting to note that both in California and Washington there are districts in which dry-farming has been practiced successfully under a precipitation of about ten inches, whereas in Utah the limit has been more nearly twelve inches.

In the Great Plains area the history of dry-farming is hopelessly lost in the greater history of the development of the eastern and more humid parts of that section of the country. The great influx of settlers on the western slope of the Great Plains area occurred in the early '80's and overflowed into eastern Colorado and Wyoming a few years later. The settlers of this region brought with them the methods of humid agriculture and because of the relatively high precipitation were not forced into the careful methods of moisture conservation that had been forced upon Utah, California, and the Columbia Basin. Consequently, more failures in dry-farming are reported from those early days in the Great Plains area than form the drier sections of the far West. Great Plains area during the later '80's. According to Payne, the crops of 189 were very good; in 1890, less so; in 1891, better; in 1892 such immense crops were raised that the settlers spoke of the sections as God's country: in 1893, there was a partial failure, and in 1894 the famous complete failure, which was followed in 1895 by a partial failure. Since that time crops have been produced annually. The dry years of 1893-1895 drove most of the discouraged settlers back to humid sections and delayed, by many years, the settlement and development of the western side of the Great Plains area. That these failures and discouragements were due almost entirely to improper methods of soil culture is very evident to the present day student of dry-farming. In fact, from the very heart of the section which was abandoned in 1893-1895 come reliable records, dating back to 1886, which show successful crop production every year.

The famous Indian Head experimental farm of Saskatchewan, at the north end of the Great Plains area has an unbroken record of good crop yields from 1888, and the early 90's were quite as dry there as farther south. However, in spite of the vicissitudes of the section, dry-farming has taken a firm hold upon the Great Plains and is now a well-established practice.

The curious thing about the development of dry-farming in Utah, California, Washington, and the Great Plains is that these four sections appear to have originated dry-farming independently of each other. True, there was considerable communication from 1849 onward between Utah and California, and there is a possibility that some of the many Utah settlers who located in California brought with them accounts of the methods of dry-farming as practiced in Utah. This, however, cannot be authenticated. It is very unlikely that time farmers of Washington learned dry-farming from their California or Utah neighbors, for until 1880 communication between Washington and the colonies in California and Utah was very difficult, though, of course, there was always the possibility of accounts of agricultural methods being carried from place to place by the moving emigrants. It is fairly certain that the Great Plains area did not draw upon the far West for dry-farm methods. The climate conditions are considerably different and the Great Plains people always considered themselves as living in a very humid country as compared with the states of the far west. It may be concluded, therefore, that there were four independent pioneers in dry-farming in United States. Moreover, hundreds, probably thousands of individual farmers over the semiarid region have practiced dry-farming thirty to fifty years with methods developed by them selves.

Although these different dry-farm sections were developed independently, yet the methods which they have finally adopted are practically identical and include deep plowing, unless the subsoil is very lifeless; fall plowing; the planting of fall grain wherever fall plowing is possible; and clean summer fallowing. About 1895 the word began to pass from mouth to mouth that probably nearly all the lands in the great arid and semiarid sections of the United States could be made to produce profitable crops without irrigation. At first it was merely a whisper; then it was talked aloud, and before long became the great topic of conservation among the thousands who love the West and wish for its development. Soon it became a National subject if discussion. Immediately after the close of the nineteenth century the new awakening had been accomplished and dry-farming was moving onward to conquer the waste places of the earth.

H. W. Campbell

The history of the new awakening in dry-farming cannot well be written without a brief account of the work of H.W. Campbell who, in the public mind, has become intimately identified with the dry-farm movement. H.W. Campbell came from Vermont to northern South Dakota in 1879, where in 1882 he harvested a banner crop, - twelve thousand bushels if wheat from three hundred acres. In 1883, on the same farm he failed completely. This experience led him to a study of the conditions under which wheat and other crops may be produced in the Great Plains area. A natural love for investigation and a dogged persistence have led him to give his life to a study of the agricultural problems of the Great Plains area. He admits that his direct inspiration came from the work of Jethro Tull, who labored two hundred years ago, and his disciples. He conceived early the idea that if the soil were packed near the bottom of the plow furrow, the moisture would be retained better and greater crop certainly would result. For this purpose the first subsurface packer was invented in 1885. Later, about 1895, when his ideas had crystallized into theories, he appeared as the publisher of Campbell's "Soil Culture and Farm Journal." One page of each issue was devoted to a succinct statement of the "Campbell Method." It was in 1898 that the doctrine of summer tillage was begun to be investigated by him.

In view of the dry-farm failures of the early '90's and the gradual dry-farm awakening of the later '90's, Campbell's work was received with much interest. He soon became identified with the efforts of the railroads to maintain demonstration farms for the benefit of intending settlers. While Campbell has long been in the service of the railroads of the semiarid region, yet it should be said in all fairness that the railroads and Mr. Campbell have had for their primary object the determination of methods whereby the farmers could be made of successful crops.

Mr. Campbell's doctrine of soil culture, based on his accumulated experience, are presented in Campbell's "Soil Culture Manual" the first edition of which appeared about 1904 and the latest edition, considerably extended, was published in 1907. The 1907 manual is the latest official word by Mr. Campbell on the principles and methods of the "Campbell system." The essential features of the system may be summarized as follows: The storage of water in the soil is imperative for the production of crops in dry years. This may be accomplished by proper tillage. Disk the lad immediately after harvest; follow as soon as possible with the plow; follow the plow with the subsurface packer; and follow the packer with the smoothing harrow. Disk the land again as early as possible in the spring and stir the soil deeply and carefully after every rain. Sow thinly in the fall with a drill. If the grain is too thick in the spring, harrow it out. To make sure of a crop, the land should be "summer tilled" which means that clean summer fallow should be practiced every other year, or as often as may be necessary.

These methods, with the exception of the subsurface packing, are sound and in harmony with the experience of the great dry-farm sections and with the principles that are being developed by scientific investigation. The "Campbell system" as it stands to-day is not the system first advocated by him. For instance, in the beginning of his work he advocated cowing grain in April and in rows so far apart that spring tooth harrows could be used for cultivating between the rows. This method, though successful in conserving moisture, is too expensive and is therefore superseded by the present methods. Moreover, his farm paper of 1896, containing a full statement of the "Campbell method," makes absolutely no mention of "summer tillage," which is now the very keystone of the system. These and other facts make it evident that Mr. Campbell has very properly modified his methods to harmonize with the best experience, but also invalidate the claim that he is the author of the dry-farm system. A weakness of the "Campbell system" is the continual insistence upon the use of the subsurface packer. As has already been shown, subsurface packing is of questionable value for successful crop production, and if valuable, the results may be much more easily and successfully obtained by the use of the disk and harrow and other similar implements now on the market. Perhaps the one great weakness in the work of Campbell is that he has not explained the principles underlying his practices. His publications only hint at the reasons. H. W. Campbell, however, has done much to popularize the subject of dry-farming and to prepare the way for others. His persistence in his work of gathering facts, writing, and speaking has done much to awaken interest in dry-farming. He has been as "a voice in the wilderness" who has done much to make possible the later and more systematic study of dry-farming. High honor should be shown him for his faith in the semiarid region, for his keen observation, and his persistence in the face of difficulties. He is justly entitled to be ranked as one of the great workers in behalf of the reclamation, without irrigation, of the rainless sections of the world.

The experiment stations

The brave pioneers who fought the relentless dryness of the Great American Desert from the memorable entrance of the Mormon pioneers into the Valley of the Great Salt Lake in 1847 were not the only ones engaged in preparing the way for the present day of great agricultural endeavor. Other, though perhaps more indirect, forces were also at work for the future development of the semiarid section. The Morrill Bill of 1862, making it possible for agricultural colleges to be created in the various states and territories, indicated the beginning of a public feeling that modern methods should be applied to the work of the farm. The passage in 187 of the Hatch Act, creating agricultural experiment stations in all of the states and territories, finally initiated a new agricultural era in the United States. With the passage of this bill, stations for the application of modern science to crop production were for the first time authorized in the regions of limited rainfall, with the exception of the station connected with the University of California, where Hilgard from 1872 had been laboring in the face of great difficulties upon the agricultural problems of the state of California. During the first few years of their existence, the stations were busy finding men and problems. The problems nearest at hand were those that had been attacked by the older stations founded under an abundant rainfall and which could not be of vital interest to arid countries.

The western stations soon began to attack their more immediate problems, and it was not long before the question of producing crops without irrigation on the great unirrigated stretches of the West was discussed among the station staffs and plans were projected for a study of the methods of conquering the desert.

The Colorado station was the first to declare its good intentions in the matter of dry-farming, by inaugurating definite experiments. By the action of the State Legislature of 1893, during the time of the great drought, a substation was established at Cheyenne Wells, near the west border of the state and within the foothills of the Great Plains area. From the summer of 194 until 190 experiments were conducted on this farm. The experiments were not based upon ay definite theory of reclamation, and consequently the work consisted largely of the comparison of varieties, when soil treatment was the all-important problem to be investigated. True, in 1898, a trial of the "Campbell method" was undertaken. By the time this station had passed its pioneer period and was ready to enter upon more systematic investigation, it was closed. Bulletin 59 of the Colorado Station, published in 1900 by J.E. Payne, gives a summary of observations made on the Cheyenne Wells substation during seven years. This bulletin is the first to deal primarily with the experimental work relating to dry-farming in the Great Plains area. It does not propose or outline any system of reclamation. Several later publications of the Colorado Station deal with the problems peculiar to the Great Plains.

At the Utah Station the possible conquest of the sagebrush deserts of the Great Basin without irrigation was a topic of common conversation during the years 1894 and 1895. In 1896 plans were presented for experiments on the principles of dry farming. Four years later these plans were carried into effect. In the summer of 1901, the author and L.A. Merrill investigated carefully the practice of the dry-farms of the state. On the basis of these observations and by the use of the established principles of the relation of water to soils and plants, a theory of dry-farming was worked out which was published in Bulletin 75 of the Utah Station in January 1902. This is probably the first systematic presentation of the principles of dry-farming. A year later the Legislature of the state of Utah made provision for the establishment and maintenance of six experimental dry-farms to investigate in different parts of the state the possibility of dry-farming and the principles underlying the art. These stations, which are still maintained, have done much to stimulate the growth of dry-farming in Utah. The credit of first undertaking and maintaining systematic experimental work in behalf of dry-farming should be assigned to the state of Utah. Since dry-farm experiments began in Utah in 1901, the subject has been a leading one in the Station and the College. A large number of men trained at the Utah Station and College have gone out as investigators of dry-farming under state and Federal direction.

The other experiment stations in the arid and semiarid region were not slow to take up the work for their respective states. Fortier and Linfield, who had spent a number of years in Utah and had become somewhat familiar with the dry-farm practice of that state, initiated dry-farm investigation in Montana, which have been prosecuted with great vigor since that time. Vernon, under the direction of Foster, who had spent four years in Utah as Director of the Utah Station, initiated the work in New Mexico. In Wyoming the experimental study of dry-farm lands began by the private enterprise of H.B. Henderson and his associates. Later V.T. Cooke was placed in charge of the work under state auspices, and the demonstration of the feasibility of dry-farming in Wyoming has been going on since about 1907. Idaho has also recently undertaken dry-farm investigations. Nevada, once looked upon as the only state in the Union incapable of producing crops without irrigation, is demonstrating by means of state appropriations that large areas there are suitable for dry-farming. In Arizona, small tracts in this sun-baked state are shown to be suitable for dry-farm lands. The Washington Station is investigating the problems of dry-farming peculiar to the Columbia Basin and the staff of the Oregon Station is carrying on similar work. In Nebraska, some very important experiments on dry-farming are being conducted. In North Dakota there were in 1910 twenty-one dry-farm demonstration farms. In South Dakota, Kansas, and Texas, provisions are similarly made for dry-farm investigations. In fact, up and down the Great Plaines area there are stations maintained by the state or Federal government for the purpose of determining the methods under which crops can be produced without irrigation.

At the head of the Great Plains area at Saskatchewan one of the oldest dry-farm stations in America is located (since 1888). In Russia several stations are devoted very largely to the problems of dry land agriculture. To be especially mentioned for the excellence of the work done are the stations at Odessa, Cherson, and Poltava. This last-named Station has been established since 1886.

In connection with the work done by the experiment stations should be mentioned the assistance given by the railroads. Many of the railroads owning land along their respective lines are greatly benefited in the selling of these lands by knowledge of the methods whereby the lands may be made productive. However, the railroads depend chiefly for their success upon the increased prosperity of the population along their lines and for the purpose of assisting the settlers in the arid West considerable sums have been expended by the railroads in cooperation with the stations for the gathering of information of value in the reclamation of arid lands without irrigation.

It is through the efforts of the experiment stations that the knowledge of the day has been reduced to a science of dry-farming. Every student of the subject admits that much is yet to be learned before the last word has been said concerning the methods of dry-farming in reclaiming the waste places of the earth. The future of dry-farming rests almost wholly upon the energy and intelligence with which the experiment stations in this and other countries of the world shall attack the special problems connected with this branch of agriculture.

The united state Department of Agriculture

The Commissioner of Agriculture of ht United States was given a secretary ship in the president's Cabinet in 1889. With this added dignity, new life was given to the department. Under the direction of J. Sterling Morton preliminary work of great importance was done. Upon the appointment of James Wilson as Secretary of Agriculture, the department fairly leaped into a fullness of organization for the investigation of the agricultural problems of the country. From the beginning of its new growth the United States Department of Agricultural has given some thought to the special problems of the semiarid region, especially that part within the Great Plains. Little consideration was at first given to the far West. The first method adopted to assist the farmers of the plains was to find plants with drought resistant properties. For that purpose explores were sent over the earth, who returned with great numbers of new plants or varieties of old plants, some of which, such as the durum wheat's, have shown themselves of great value in American Agriculture. The Bureaus of plant Industry, Soils, Weather and Chemistry have all from the first given considerable attention to the problems of the arid region. The Weather Bureau, long established and with perfected methods, has been invaluable in guiding investigators into regions where experiments could be undertaken with some hope of success. The Department of Agriculture was somewhat slow, however, in recognizing dry-farming as a system of agriculture requiring special investigation. The final recognition of the subject came with the appointment, in 1905 of Chilcott as expert in charge of dry-land investigations. At the present time an office of dry-land in investigations had been established under the Bureau of plant Industry, which cooperates with a number of other divisions of the Bureau in the investigation of the conditions and methods of dry-farming. A large number of stations are maintained by the Department over the arid and semiarid area for the purpose of studying special problems, many of which are maintained in connection with the state experiment stations. Nearly all the department experts engaged in dry-farm investigation have been drawn from the service of the state stations and in these stations had received their special training for their work. The United States Department of Agriculture has chosen to adopt a strong conservation in the matter of dry farming. It may be wise for the Department, as the official head of the agricultural interests of the country, to use extremes care in advocating the settlement of a region in which, in the past, farmers had failed to make a living, yet this conservatism has tended to hinder the advancement of dry-farming and has placed the departmental investigations of dry-farming in point of time behind the pioneer investigations of the subject.

The Dry-farming Congress

As the great dry-farm wave swept over the country, the need was felt on the part of experts and laymen of some means whereby dry-farm ideas from all parts of the country could be exchanged. Private individuals by the thousands and numerous state and governmental stations were working separately and seldom had a chance of comparing notes and discussing problems. A need was felt for some central dry-farm organization. An attempt to fill this need was made by the people of Denver, Colorado, when Governor Jesse F. Me Donald of Colorado issued a call for the firs Dry-farming Congress to be held in Denver, January 24, 25, and 26, 1907. These dates were those of the annual stock show which had become permanent institutions of Denver and, in fact, some of those who were instrumental in the calling of the Dry-farming Congress thought that it was a good scheme to bring more people to the stock show. To the surprise of many the Dry-farming Congress became the lead in feature of the week. Representative were present from practically all the states interested in dry-farming and from some of the humid states. Utah, the pioneer dry-farm state, was represented by a delegation second in size only to that of Colorado, where the Congress was held. The call for this Congress was inspired, in part at least, by real estate men. Who saw in the dry-farm movement an opportunity to relive themselves of large areas of cheap land at fairly good prices? The Congress proved, however, to be a businesslike meeting which took hold of the questions in earnest, and form the very first made it clear that the real estate agent was not a welcome member unless he came with perfectly honest methods.

The second dry-farming Congress was held January 22 to 25, 1908, in Salt lake city, Utah, under the presidency of Fisher Harris. It was even better attended than the first. The proceeding shows that it was a Congress at which the dry-farm experts of the country stated their findings. A large exhibit of dry-farm products was held in connection with this Congress, where ocular demonstrations of the possibility of dry-farming were given any doubting Thomas.

The third Dry-farming Congress was held February 23 to 25, 1909, at Cheyenne, Wyoming, under the presidency of Governor W.W. Brooks of Wyoming. An unusually severe snowstorm preceded the Congress, which prevented many from attending, yet the number present exceeded that at any of the preceding Congresses. This congress was made notable by the number of foreign delegates who had been sent by their respective countries to investigate the methods pursued in America for the reclamation of the arid districts. Among these delegates were representatives from Canada, Australia, The Transvaal, Brazil, and Russia.

The fourth Congress was held October 26 to 28, 1909, in Billings, Montana, under the presidency of Governor Edwin L. Morris of Montana. The uncertain weather of the winter months had led the previous congress to adopt a time in the autumn as the date of the annual meeting. This Congress became a session at which many of the principles discussed during the three preceding Congresses were crystallized into definite statements and agreed upon by workers from various parts of the country. A number of foreign representatives were present again. The problems of the Northwest and Canada were given special attention. The attendance was larger than at any of the preceding Congresses.

The fifth congress will be held under the presidency of Hon. F.W. Mondell of Wyoming at Spokane, Washington, during October, 1910. It promises to exceed any preceding congress in attendance and interest.

The Dry-farming Congress has made itself one of the most important factors in the development of methods for the reclamation of the desert. Its published reports are the most valuable publications dealing with dry-land agriculture. Only simple justice is done when it is stated that the success of the Dry-farming Congress is due in a large measure to the untiring and intelligent efforts of John T. Burns, who is the permanent secretary of the congress, and who was a member of the first executive committee.

Nearly all the arid and semiarid states have organized state dry-farming congresses. The first of these was the Utah Dry-farming Congress, organized about two months after the first congress held in Denever. The president is L.A. Merrill, one of the pioneer dry-farm investigators of the Rockies.

Jethro Tull (see frontispiece)

A sketch of the history of dry-farming would be incomplete without a mention of the life and work of Jethro Tull. The agricultural doctrines of this man, interpreted in the light of modern science, are those, which underlie modern dry-farming. Jethro Tull was born in Berkshire, England, 1674, and died in 1741. He was a lawyer by profession, but his health was so poor that he could not practice his profession of a quiet farm. His life work was done in the face of great physical sufferings. In spite of physical infirmities, he produced a system of agriculture which, viewed in the light of our modern knowledge is little short of marvelous. The chief inspiration of his system came from a visit paid to south of France, where he observed "near Fronting and Setts, Languedoc" that the vineyards were carefully plowed and tilled in order to produce the largest crops of the best grapes. Upon the basis of this observation he instituted experiments upon his own farms and finally developed his system, which may be summarized as follows: The amount of seed to be used should be proportional to the condition of the land, especially to the moisture that is in it. To sown by drill methods. Tull, as has already been observed was the inventor of the seed drill which is now a feature of all modern agriculture. Plowing should be done deeply and frequently; two plowing for one crop would do no injury and frequently would result in an increased yield. Finally, as the most important [principles of the system, the soil should be cultivated continually the argument being that by continuous cultivation the fertility of ht soil would be increased the water would be conserved and as the soil became more fertile less water would be used. To accomplish such rather far apart, so far indeed that a horse carrying a cultivator could walk between them. The horse-hoeing idea of the system became fundamental and gave the name to his famous book, "The Horse Hoeing Husbandry," by Jethro Tull, published in pats from 1731 to 1741. Tull held that the soil between the rows was essentially being fallowed and that the next year the seed could be planted between the rows of the preceding year and in that way the fertility could be maintained almost indefinitely. If this method were not followed half of the soil could lie fallow every other year and be subjected to continuous cultivation. Weeds consume water and fertility and, therefore, fallowing and all the culture must be perfectly clean. To maintain fertility a rotation of crops grain crop; turnips the root crop; and alfalfa a very desirable crop.

It may be observed that these teachings are sound and in harmony with the best knowledge of to-day and that they are the very practice which is now being advocated in all dry-farm sections. This is doubly curious because Tull lived in a humid country. However, if may be mentioned that his farm consisted of a very poor chalk soil, so that the conditions under which he labored were more nearly those of an arid country than could ordinarily be found in a country of abundant rainfall. While the practices of Jethro Tull were in themselves very good and in general can be adopted to-day, yet his interpretation of the limited knowledge of his day, this was only to be expected. For instance, he believed so thoroughly in the value of cultivation of the soil, that he thought it would take the place of all other methods of maintaining soil-fertility. In fact, he declared dictionary that "tillage is manure," which we are very certain at this time is fallacious. Jethro Tull is one of there great investigators of the world. In recognition of the fact that, though living two hundred years ago in a humid country, he was able to develop the fundamental practices of soil culture now used in dry-farming, the honor has been done his memory of placing his portrait as the frontispiece of his volume.

CHAPTER XVIII

THE PRESENT STATUS OF DRY-FARMING

It is difficult to obtain a correct view of the present status of dry-farming, first, because dry-farm surveys are only beginning to be made and, secondly, because the area under dry-farm cultivation is increasing daily by leaps and bounds. All arid and semiarid parts of the world are reaching out after methods of soil culture whereby profitable crops may be produced without irrigation, and the practice of dry-farming, according to modern methods, is now followed in many diverse countries. The United States undoubtedly leads at present in the area actually under dry-farming, but, in view of the immense dry-farm districts in other parts of the world, it is doubtful if the United States will always maintain its supremacy in dry-farm acreage. The leadership in the development of a science of dry-farming will probably remain with the United States for years, since the numerous experiment stations established for the study of the problems of farming without irrigation have their work well under way, while, with the exception of one or two station in Russia and

Canada, no other countries have experiment stations for the study of dry-farming in full operation. The reports of the Dry-farming Congress furnish practically the only general information as to the status of dry-farming in the states and territories of the United States and in the countries of the world.

California

In the state of the California dry-farming has been firmly established for more than a generation. The chief crop of the California dry-farms is wheat, though the other grains, root crops, and vegetables are also grown without irrigation under a comparatively small rainfall. The chief dry-farm areas are found in the Sacramento and the San Joaquin valleys. In the Sacramento Valley the precipitation is fairly large, but in the San Joaquin Valley it is very small. Some of the most successful dry-farms of California have produced well for a long succession of years under a rainfall of ten inches and less. California offers a splendid example of the great danger that besets all dry-farm sections. For a generation wheat has been produced on the fertile Californian soils without manuring of any kind. As a consequence, the fertility of the soils has been so far depleted that at present it is difficult to obtain paying crops without irrigation on soils that formerly yielded bountifully. The living problem of the dry-farms in California is the restoration of the fertility which has been removed from the soils by unwise cropping. All other dry-farm districts should take to heart this lesson, for, though crops may be produced on fertile soils for one, two, or even three generations without manuring, yet the time will come when plant-food must be added to the soil in return for that which has been removed by the crops. Meanwhile, California offers, also, an excellent example of the possibility of successful dry-farming through long periods and under varying climatic conditions. In the Golden State dry-farming is a fully established practice; it has long since passed the experimental stage.

Columbia River Basin

The Columbia River Basin includes the state of Washington, most of Oregon, the northern and central part of the Idaho, western Montana, and extends into British Columbia. It includes the section often called the Inland Empire, which alone covers some one hundred and fifty thousand square miles. The chief dry-farm crop of this region is wheat; in fact, western Washington or the "Palouse country" is famous for its wheat-producing powers. The other grains, potatoes, roots, and vegetables are also grown without irrigation. In the parts of this dry-farm district where the rainfall is the highest, fruits of many kinds and of a high quality are grown without irrigation. It is estimated that at least two million acres are being dry-farmed in this district. Dry-farming is fully established in the Columbia River Basin. One farmer is reported to have raised in one year on his own farm two hundred and fifty thousand bushels of wheat. In one section of the district where the rainfall for the last few years has been only about ten or eleven inches, wheat has been produced successfully. This corroborates the experience of California, that wheat may really be grown in localities where the annual rainfall is not above ten inches. The most modern methods of dry-farming are followed by the farmers of the Columbia River Basin, but little attention has been given to soil-fertility, since soils that have been farmed for a generation still appear to retain their high productive powers. Undoubtedly, however, in this district, as in California, the question of soil-fertility will be an important one in the near future. This is one of the great dry-farm districts of the world.

The Great Basin

The Great Basin includes Nevada, the western half of Utah, a small part of southern Oregon and Idaho, and also a part of Southern California. It is a great interior basin with all its rivers draining into salt lakes or dry sinks. In recent geological times the Great Basin was filled with water, forming the great Lake Bonneville which drained into the Columbia River. In fact, the Great Basin is made up of a series of great valleys, with very level floors, representing the old lake bottom. On the bench lands are seen, in many places the effects of the wave action of the ancient lake. The chief dry-farm crop of this district is wheat, but the other grains, including corn, are also produced successfully. Other crops have been tried with fair success, but not on a commercial scale. Grapevines have been made to grow quite successfully without irrigation on the bench lands. Several small orchards bearing luscious fruit are growing on the deep soils of the Great Basin without the artificial application of water. Though the first dry-farming by modern peoples was probably practiced in the Great Basin, yet the area at present under cultivation is not large, possibly a little more than four hundred thousand acres. Dry-farming, however, is well established. There are large areas, especially in Nevada, that receive less than ten inches of rainfall annually, and one of the leading problems before the dry-farmers of this district is the determination of the possibility of producing crops upon such lands without irrigation. On the older dry-farms, which have existed in some cases from forty to fifty years, there are no signs of diminution of soil-fertility. Undoubtedly, however, even under the conditions of extremely high fertility prevailing in the Great Basin, the time will soon come when the dry-farmer must make provision for restoring to the soil some of the fertility taken away by crops. There are millions of acres in the Great Basin yet to be taken up and subjected to the will of the dry-farmer.

Colorado and Rio Grande River Basins

The Colorado and Rio Grande River Basins include Arizona and the western part of New Mexico. The chief dry-farm crops of this dry district are wheat, corn, and beans. Other crops have also been grown in small quantities and with some success. The area suitable for dry-farming in this district has not yet been fully determined and, therefore, the Arizona and New Mexico stations are undertaking dry-farm surveys of their respective stats. In spite of the fact that Arizona is generally looked upon as one of the driest states of the Union, dry-farming is making considerable headway there. In New Mexico, five sixths of all the homestead applications during the last year were for dry-farm lands; and, in fact, there are several prosperous communities in New Mexico which are subsisting almost wholly on dry-farming. It is only fair to say, however, that dry-farming is not yet well established in this district, but that the prospects are that the application of scientific principles will soon make it possible to produce profitable crops without irrigation in large parts of the Colorado and Rio Grande River Basins.

The mountain states

This district includes a part of Montana, nearly the whole of Wyoming and Colorado, and part of eastern Idaho. It is located along the backbone of the Rocky Mountains. The farms are located chiefly in valleys and on large rolling table-lands. The chief dry-farm crop is wheat, though the other crops with are grown elsewhere on dry-farms may be grown here also. In Montana there is a very large area of land which has been demonstrated to be well adapted for dry-farm purposes. In Wyoming, especially on the eastern as well as on the far western side, dry-farming has been shown to be successful, but the area covered at the present time is comparatively small. In Idaho, dry-farming is fairly well established. In Colorado, likewise, the practice is very well established and the area is tolerably large. All in all, throughout the mountain states dry-farming maybe said to be well established, though there is a great opportunity for the extension of the practice. The sparse population of the western states naturally makes it impossible for more than a small fraction of the land to be properly cultivated.

The Great Plains Area

This area includes parts of Montana, North Dakota, South Dakota, Nebraska, Kansas, Wyoming, Colorado, New Mexico, Oklahoma, and Texas. It is the largest area of dry-farm land under approximately uniform conditions. Its drainage is into the Mississippi, and it covers an area of not less than four hundred thousand square miles. Dry-farm crops grow well over the whole area; in fact, dry-farming is well established in this district. In spite of the failures so widely advertised during the dry season of 1894, the farmers who remained on their farms and since that time have employed modern methods have secured wealth from their labors. The important question before the farmers of this district is that of methods for securing the best results. From the Dakotas to Texas the farmers bear the testimony that wherever the soil has been treated right, according to approved methods, there have been no crop failures.

Canada

Dry-farming has been pushed vigorously in the semiarid portions of Canada, and with great success. Dry-farming is now reclaiming large areas of formerly worthless land, especially in Alberta,

Saskatehewan; and the adjoining provinces. Dry-farming is comparatively recent in Canada, yet here and there are semiarid localities where crops have been raised without irrigation for upwards of a quarter of a century. In Alberta and other places it has been now practiced successfully for eight or ten years, and it may be said that dry-farming is a well-established practice in the semiarid regions of the Dominion of Canada.

Mexico

In Mexico, likewise, dry-farming has been tried and found to be successful. The natives of Mexico have practiced farming without irrigation for centuries; and modern methods are now being applied in the zone midway between the extremely dry and the extremely humid portions. The irregular distribution of the precipitation, the late spring and early fall frosts, and the fierce winds combine to make the dry-farm problem somewhat difficult, yet the prospects are that, with government assistance, dry-farming in the near future will become an established practice in Mexico. In the opinion of the best students of Mexico it is the only method of agriculture that can be made to reclaim a very large portion of the country.

Brazil

Brazil, which is greater in area than the United States, also has a large arid and semiarid territory which can be reclaimed only by dry-farm methods. Through the activity of leading citizens experiments in behalf of the dry-farm movement have already been ordered. The dry-farm district of Brazil receives an annual precipitation of about twenty-five inches, but irregularly distributed and under a tropical sun. In the opinion of those who are familiar with the conditions, the methods of dry-farming may be so adapted as to make dry-farming successful in Brazil.

Australia

Australia, larger than the continental United States, is vitally interested in dry-farming for one third of its vast area is under a rainfall of less than ten inches, and another third is under a rainfall of between ten and twenty inches. Two thirds of the area of Australia, if reclaimed at all, must be reclaimed by dry-farming. The realization of this condition has led several Australians to visit the United States for the purpose of learning the methods employed in dry-farming. The reports on dry-farming in America by Surveyor-General Straw-bridge and Senator J. H. McColl have done much to initiate a vigorous propaganda in behalf of dry-farming in Australia. Investigation has shown that occasional farmers are found in Australia, as in America, who have discovered for themselves many of the methods of dry-farming and have succeeded in producing crops profitably. Undoubtedly, in time, Australia will be one of the great dry-farming countries of the world.

Africa

Up to the present, South Africa only has taken an active interest in the dry-farm movement, due to the enthusiastic labors of Dr. William Macdonald of the Transvaal. The Transvaal has an average annual precipitation of twenty-three inches, with a large district that receives between thirteen and twenty inches. The rain comes in the summer, making the conditions similar to those of the Great Plains. The success of dry-farming has already been practically demonstrated. The question before the Transvaal farmers is the determination of the best application of water conserving methods under the prevailing conditions. Under proper leadership the Transvaal and other portions of Africa will probably join the ranks of the larger dry-farming countries of the world.

Russia

More than one fourth of the whole of Russia is so dry as t be reclaimable only by dry-farming. The arid area of Southern European Russia has a climate very much like that of the Great Plains. Turkestan and middle Asiatic Russia have a climate more like that of the Great Basin. In a great number of localities in both European and Asiatic Russia dry-farming has been practiced for a number of years. The methods employed have not been of the most refined kind, due, possibly, to the condition of the people constituting the farming class. The government is now becoming interested in the matter and there is no doubt that dry-farming will also be practiced on a very large scale in Russia.

Turkey

Turkey has also a large area of arid land and, due to American assistance, experiments in dry-farming are being carried on in various parts of the country. It is interesting to learn that the experiments there, up to date, have been eminently successful and that the prospects now are that modern dry-farming will soon be conducted on a large scale in the Ottoman Empire.

Palestine

The whole of Palestine is essentially arid and semi-arid and dry-farming there has been practiced for centuries. With the application of modern methods it should be more successful than ever before. Dr. Aaronsohn states that the original wild wheat from which the present varieties of wheat have descended has been discovered to e a native of Palestine.

China

China is also interested in dry-farming. The climate of the drier portions of China is much like that of the Dakotas. Dry-farming there is of high antiquity, though, of course, the methods are not those that have been developed in recent years. Under the influence of the more modern methods dry-farming should spread extensively throughout China and become a great source of profit to the empire. The results of dry-farming in China are among the best.

These centuries have been mentioned simply because they have been represented at the recent Dry-farming Congresses. Nearly all of the great countries of the world having extensive semiarid areas are directly interested in dry-farming. The map on pages 30 and 31 shows that more than 55 per cent of the world's surface receives an annual rainfall of less than twenty inches. Dry-farming is a world problem and as such is being received by the nations.

CHAPTER XIX

THE YEAR OF DROUGHT

The Shadow of the year of Drought still obscures the hope of many a dry-farmer. From the magazine page and the public platform the prophet of evil, thinking himself a friend of humanity, solemnly warns against the arid region and dry-farming, for the year of drought, he says, is sure to come again and then will be repeated the disasters of 1893-1895. Beware of the year of drought. Even successful dry-farmers who have obtained good crops every year for a generation or more are half led to expect a dry year – one so dry that crops will fail in spite of all human effort. The question is continually asked, "Can crop yields reasonably be expected every year, through a succession of dry years, under semiarid conditions, if the best methods of dry-farming be practiced?" In answering this question, it may be said at the very beginning, that when the year of drought is mentioned in connection with dry-farming, sad reference is always made to the experience on the Great Plains in the early years of the 90's. Now the fact of the matter is, that while the years of 1893, 1894, and 1895 were dry years, the only complete failure came in 1894. In spite of the improper methods practiced by the settlers, the willing soil failed to yield a crop only one year. Moreover, it should not be forgotten that hundreds of farmers in the driest section during this dry period, who instinctively or otherwise farmed more nearly right, obtained good crops even in 1894. The simple practice of summer fallowing, had it been practiced the year before, would have insured satisfactory crops in the driest year. Further, the settlers who did not take to their heels upon the arrival of the dry year are still living in large numbers on their homesteads and in numerous instances have accumulated comfortable fortunes from the land which has been held up so long as a warning against settlement beyond a humid climate. The failure of 1894 was due as much to a lack of proper agricultural information and practice as to the occurrence of a dry year.

Next, the statement is carelessly made that the recent success in dry-farming is due to the fact that we are now living in a cycle of wet years, but that as soon as the cycle of dry years strikes the country dry-farming will vanish as a dismal failure. Then, again, the theory is proposed that the climate is permanently changing toward wetness or dryness and the past has no meaning in reading the riddle of the future. It is doubtless true that no man may safely predict the weather for future generations; yet, so far as human knowledge goes, there is no perceptible average change in the climate from period to period within historical time; neither are there protracted dry periods followed by protracted wet periods. The fact is, dry and wet years may alternate with a succession of somewhat dry years, but the average precipitation from decade to decade is very nearly the same. True, there will always be a dry year, that is, the driest year of a series of years, and this is the supposedly fearful and fateful year of drought. The business of the dry-farmer is always to farm so as to be prepared for this driest year whenever it comes. If this be done, the farmer will always have a crop: in the wet years his crop will be large; in the driest year it will be sufficient to sustain him.

So persistent is the half-expressed fear that this driest year makes it impossible to rely upon dry-farming as a permanent system of agriculture that a search has been made for reliable long records of the production of crops in arid and semiarid regions. Public statements have been made by many perfectly reliable men to the effect that crops have been produced in diverse sections over long periods of years, some as long as thirty-five or forty years, without one failure having occurred. Most of these statements, however, have been general in their nature and not accompanied by the exact yields from year to year. Only three satisfactory records have been found in a somewhat careful search. Others no doubt exist.

The Barnes farm

The first record was made by Senator J. G. M. Barnes of Kaysville, Utah. Kaysville is located in the Great

Salt Lake Valley, about fifteen miles north of Salt Lake City. The climate is semiarid; the precipitation comes mainly in the winter and early spring; the summers are dry, and the evaporation is large. Senator Barnes purchased ninety acres of land in the spring of 1887 and had it farmed under his own supervision until 1906. he is engaged in commercial enterprises and did not, himself, do any of the work on the farm, but employed men to do the necessary labor. However, he kept a close supervision of the farm and decided upon the practices which should be followed. From seventy-eight to eighty-nine acres were harvested for each crop, with the exception of 1902, when all but about twenty acres was fired by sparks from the passing railroad train. The plowing, harrowing, and weeding were done very carefully. The complete record of the Barnes dry-farm from 1887 to 1905 is shown in the table 21.

The first plowing was given the farm in May of 1887, and, with the exception of 1902, the land was invariably plowed in the spring. With fall plowing the yields would undoubtedly have been better. The first sowing was made in the fall of 1887, and fall grain was grown during the whole period of observation. The seed sown in the fall of 1887 came up well, but was winter-killed. This is ascribed by Senator Barnes to the very dry winter, though it is probable that the soil was not sufficiently well stored with moisture to carry the crop through. The farm was plowed again in the spring of 1888, and another crop sown in September of the same year. In the summer of 1889, 22½ bushels of wheat were harvested to the acre.

Encouraged by this good crop Mr. Barnes allowed a volunteer crop to grow that fall and the next summer harvested as a result 15½ bushels of wheat to the acre. The table shows that only one crop smaller than this was harvested during the whole period of nineteen years, namely, in 1903, when the same thing was done, and one crop was made to follow another without an intervening fallow period. This observation is an evidence in favor of clean summer fallowing. The largest crop obtained, 28.9 bushels per acre in 1902, was gathered in a year when the next to the lowest rainfall of the whole period occurred, namely, 11.41 inches.

TABLE: 21
RECORD OF THE BARNES DRY-FARM, SALT LAKE VALLEY, UTAH (90 ACRES)

Year	Annual rainfall (Inches)	Yield per acre (Bu.)	When plowed	When sown
1887	11.66	-	May	Sept.
1888	13.62	Failure	May	Sept.
1889	18.46	22.5	-	Volunteer[1]
1890	10.33	15.2	-	-
1891	15.92	Fallow	May	Fall
1892	14.08	19.3	-	-
1893	17.35	Fallow	May	Fall
1894	15.27	26.0	-	-
1895	11.95	Fallow	May	Aug.
1896	18.42	22.0	-	-

[1] *About four acres were sown on stubble.*

Year	Annual rainfall (Inches)	Yield per acre (Bu.)	When plowed	When sown
1897	16.74	Fallow	Spring	Fall
1898	16.09	26.0	-	-
1899	17.57	Fallow	May	Fall
1900	11.53	23.5	-	-
1901	16.08	Fallow	Spring	Fall
1902	11.41	28.9	Sept.	Fall
1903	14.62	12.5	-	-
1904	16.31	Fallow	Spring	Fall
1905	14.23	25.8	-	-

The precipitation varied during the nineteen years from 10.33 inches to 18.46 inches. The variation in yield per acre was considerably less than this, not counting the two crops that were grown immediately after another crop. All in all, the unique record of the Barnes dry-farm shows that through a period of nineteen years, including dry and comparatively wet years, there was absolutely no sign of failure, except in the first year, when probably the soil had not been put in proper condition to support crops. In passing it may be mentioned that, according to the records furnished by Senator Barnes, the total cost of operating the farm during the nineteen years was $4887.69; the total income was $10,144.83. The difference, $5257.14, is a very fair profit on the investment of $1800 – the original cost of the farm.

THE INDIAN HEAD FARM

AN equally instructive record is furnished by the experimental farm located at Indian Head in

Saskatchewan, Canada, in the northern part of the Great Plains area. According to Always, the country is in appearance very much like western Nebraska and Kansas; the climate is distinctly arid, and the precipitation comes mainly in the spring and summer. It is the only experimental dry-farm in the Great Plains area with records that go back before the dry years of early 90's. In 1882 the soil of this farm was broken, and it was farmed continuously until 1888, when it was made an experimental farm under government supervision. The following table shows the yields obtained from the year 1891, when the precipitation records were first kept, to 1909:-

TABLE: 22
RECORD OF INDIAN HEAD EXPERIMENTAL FARM AND MOTHERWELL'S FARM, SASKATCHEWAN, CANADA

Year	Annual rainfall (inches)[2]	Bushels of Wheat per acre Experimental farm		Bushels of Wheat per acre Motherwell's farm
		Fallow	Stubble	
1891	14.03	35	32	30
1892	6.92	28	21	28
1893	10.11	35	22	34
1894	3.90	17	9	24
1895	12.28	41	22	26
1896	10.59	39	29	31
1897	14.62	33	26	35
1898	14.03	32	-	27
1899	9.44	33	-	33

[2] *Snowfall not included. This has varied from 2.3 to 1.3 inches of water.*

Year	Annual rainfall (inches)[2]	Bushels of Wheat per acre Experimental farm		Bushels of Wheat per acre Motherwell's farm
		Fallow	Stubble	
1900	11.74	17	5	25
1901	20.22	49	38	51
1902	10.73	38	22	28
1903	15.55	35	15	31
1904	11.96	40	29	35
1905	19.17	42	18	36
1906	13.21	26	13	38
1907	15.03	18	18	15
1908	13.17	29	14	16
1909	13.96	28	15	23
Average	-	32.4	20.5	-

The annual rainfall shown in the second column does not include the water which fell in the form of snow. According to the records at hand, the annual snow fall varied from 2.3 to 1.3 inches of water, which should be added to the rainfall given in the table.

Even with this addition the rainfall shows the district to be of a distinctly semiarid character. It will be observed that the precipitation varied from 3.9 to 20.00 inches, and that during the early 90's several rather dry years occurred. In spite of this large variation good crops have been obtained during the whole period of nineteen years. Not one failure is recorded. The lowest yield of 17 bushels per acre came during the very dry year of 1894 and during the somewhat dry year of 1900. some of the largest yield were obtained in seasons when the rainfall was only near the average. As a record showing that the year of drought need not be feared when dry-farming is done right, this table is of very high interest. It may be noted, incidentally, that throughout the whole period wheat following a fallow always yielded higher than wheat following the stubble. For the nineteen years, the difference was as 32.4 bushels is to 20.5 bushels.

The motherwell farm

In the last column of the table are shown the annual yields of wheat obtained on the farm of Commissioner Motherwell of the province of Saskatchewan. This private farm is located some twenty-five miles away from Indian Head, and the rainfall records of the experimental farm are, therefore, only approximately accurate for the Motherwell farm. The results on this farm may well be compared to the Barnes results of Utah, since they were obtained on a private farm. During the period of nineteen years good crops were invariably obtained; even during the very dry year of 1894, a yield of twenty-four bushels of wheat to the acre was obtained. Curiously enough, the lowest yields of fifteen and sixteen bushels to the acre were obtained in 1907 and 1908 when the precipitation was fairly good and must be ascribed to some other factor than that of precipitation. The record of this farm shows conclusively that with proper farming there is no need to fear the year of drought.

The Utah drought of 1910

During the year of 1910 only 2.7 niches of rain fell in Salt Lake City from March 1 to the July harvest, and all of this in March, as against 7.18 inches during the same period the preceding year. In other parts of the state much less rain fell; in fact, in the southern part of the state the last rain fell during the last week of December, 1909. The drought remained unbroken until long after the wheat harvests. Great fear was expressed that the dry-farms could not survive so protracted a period of drought. Agents, sent out over the various dry-farm districts, reported late in June that wherever clean summer fallowing had been practiced the crops were in excellent condition; but that wherever careless methods had been practiced, the crops were poor or killed. The reports of the harvest in July of 1910 showed that fully 85 per cent of an average crop was obtained in spite of the protracted drought wherever the soil came into the spring well stored with moisture, and in many instances full crops were obtained.

Over the whole of the dry-farm territory of the United States similar conditions of drought occurred. After the harvest, however, every state reported that the crops were well up to the average wherever correct methods of culture had been employed.

These well-authenticated records from true semi-arid districts, covering the two chief types of winter and summer precipitation, prove that the year of drought, or the driest year in a twenty-year period, does not disturb agricultural conditions seriously in localities where the average annual precipitation is not too low, and where proper cultural methods are followed. That dry-farming is a system of agricultural practice which requires the application of high skill and intelligence is admitted; that it is precarious is denied. The year of drought is ordinarily the year in which the man failed to do properly his share of the work.

CHAPTER XX
DRY-FARMING IN A NUTSHELL

- Locate the dry-farm in a section with an annual precipitation of more than ten inches and if possible, with small wind movement. One man with four horses and plenty of machinery cannot handle more than from 160 to 200 acres. Farm fewer acres and farm them better.

- Select a clay loam soil. Other soils may be equally productive, but are cultivated properly with somewhat more difficulty.

- Make sure, with the help of the soil auger, that the soil is of uniform structure to a depth of at least eight feet. If streaks of loose gravel or layers of hardpan are near the surface, water may be lost to the plant roots.

- After the land has been cleared and broken let it lie fallow with clean cultivation, for one year. The increase in the first and later crops will pay for the waiting.

- Always plow the land early in the fall, unless abundant experience shows that fall plowing is an unwise practice in the locality. Always plow deeply unless the subsoil is infertile, in which case plow a little deeper each year until eight or ten inches are reached. Plow at least once for each crop. Spring plowing, if practiced, should be done as early as possible in the season.

- Follow the plow, whether in the fall or spring, with the disk and that with the smoothing harrow, if crops are to be sown soon afterwards. If the land plowed in the fall is to lie fallow for the winter, leave it in the rough condition, except in localities where there is little or not snow and the winter temperature is high.

- Always disk the land in early spring, to prevent evaporation. Follow the disk with the harrow. Harrow, or in some other way stir the surface of the soil after every rain. If crops are on the land, harrow as long as the plants will stand it. If hoed crops, like corn or potatoes, are grown, use the cultivator throughout the season. A deep mulch or dry soil should cover the land as far as possible throughout the summer. Immediately after harvest disk the soil thoroughly.

- Destroy weeds as soon as they show themselves. A weedy dry-farm is doomed to failure.

- Give the land an occasional rest, that is, a clean summer fallow. Under a rainfall of less than fifteen inches, the land should be summer fallowed every other year; under an annual rainfall of fifteen to twenty inches, the summer fallow should occur every third or fourth year. Where the rainfall comes chiefly in the summer, the summer fallow is less important in ordinary years than where the summers are dry and the winters wet. Only an absolutely clean fallow should be permitted.

- The fertility of dry-farm soils must be maintained. Return the manure; plow under green leguminous crops occasionally and practice rotation. On fertile soils plants mature with the least water.

- Sow only by the drill method. Wherever possible use fall varieties of crops. Plant deeply – three or four inches for grain. Plant early in the fall, especially if the land has been summer fallowed. Use only about one half as much seed as is recommended for humid-farming.

- All the ordinary crops may be grown by dry-farming. Secure seed that has been raised on dry-farms. Look out for new varieties, especially adapted for dry-farming, that maybe brought in. Wheat is king in dry-farming; corn a close second. Turkey wheat promises the best.

- Stock the dry-farm with the best modern machinery. Dry-farming is possible only because of the modern plow, the disk, the drill seeder, the harvester, the header, and the thresher.

- Make a home on the dry-farm. Store the flood waters in a reservoir, or pump the underground waters, for irrigating the family garden. Set out trees, plant flowers and keep some live stock.

- Learn to understand the reasons back of the principles of dry-farming, apply the knowledge vigorously and the crop cannot fail.
- Always farm as if a year of drought were coming.
- Man, by his intelligence, compels the laws of nature to do his bidding, and thus he achieves joy.
- "And God blessed them – and God said unto them, Be fruitful and multiply and replenish the earth, *and subdue it*".

APPENDIX A

A PARTIAL BIBLIOGRAPHY OF THE LITERATURE OF DRY-FARMING

1897

- Some Interesting Soil Problems. Milton Whitney. Yearbook, U. S. Department of Agriculture for 1897, page 429.

1900

- The Plains. J. E. Payne. Bulletin 59, Colorado Experiment Station.
- Successful Wheat Growing in Semiarid Districts. M. A. Carleton. Yearbook, U. S. Department of Agriculture for 1900, page 529.

1901

- Macaroni Wheats. M. A. Carleton. Bulletin No. 3, Bureau of Plant Industry, U. S. Department of Agriculture.
- Emmer: A Grain for the Semiarid Regions. M. A. Carleton. Farmers' Bulletin 139, U. S. Department of Agriculture.

1902

- Arid-farming or Dry-farming. J. A. Widtsoe and L. A. Merrill. Bulletin 75, Utah Experiment Station.
- The Algerian Durum Wheats: A Classified List, with Descriptions. C. S. Scofield. Bulletin No. 7, Bureau of Plant Industry, U. S. Department of Agriculture.

1903

- Investigation of the Great Plains and Unirrigated Lands of Eastern Colorado: Seven Years Study. J. E. Payne. Bulletin 77, Colorado Experiment Station.
- The Description of Wheat Varieties. C. S. Scofield. Bulletin No. 47, Bureau of Plant Industry, U. S Department of Agriculture.

1905

- Macaroni Wheats. T. L. Lyon. Bulletin No. 78, Nebraska Experiment Station.
- Report of the Edgley Substation. North Dakota Experiment Station.
- The Thorough Tillage System for the Plains of Colorado Experiment Station.
- Agriculture without Irrigation in the Sahara Desert. T. H. Kearney. Bulletin No. 86, Bureau of Plant Industry, U. S. Department of Agriculture.

- Macaroni Wheat: Its Milling and Chemical Characteristics and Its Adaptation for Making Bread and Macaroni. J. H. Shepherd. Bulletin No. 92, South Dakota Experiment Station.
- The Relation of Irrigation to Dry-farming. El-wood Mead. Yearbook, U. S. Department of Agriculture for 1905, page 423.
- Arid-farming in Utah: First Report of the State Experimental Arid Farms. J. A. Widtsoe and L. A. Merrill. Bulletin No. 91, Utah Experiment Station.

1906

- Agriculture and Irrigation. J. H. McColl, M. P. Bendigo, Australia.
- Management of Soils to Conserve Moisture with Special Reference to Semiarid Conditions. George H. Failyer. Farmers Bulletin No. 266, U. S. Department of Agriculture.
- Saccharine Sorghums for Forage. C. R. Ball. Farmers' Bulletin No. 246, U. S. Department of Agriculture.
- Arid-farming Investigations. W. M. Jardine. Bulletin 100, Utah Experiment Station.
- Macaroni or Durum Wheats. J. H. Shepherd. Bulletin 99, South Dakota Experiment Station.

1907

- Dry Land Farming in the Great Plains Area. E. C. Chilcott. Yearbook, U. S. Department of Agriculture for 1907, page 451.
- Dry-farming in the Great Basin. C. S. Scofield. Bulletin 103, Bureau of Plant Industry, U. S. Department of Agriculture.
- Dry-farming in New Mexico. J. J. Vernon. Bulletin No. 61, New Mexico Experiment Station.
- The Culture and Uses of Brome Grass. R. A. Oakley. Bulletin 111, Part V, Bureau of Plant Industry, U. S. Department of Agriculture.
- Farm Practice in the Columbia Basin Uplands. Byron Hunter. Farmers' Bulletin 295, U. S. Department of Agriculture.
- Dry-farming in Montana. F. B. Linfield and Alfred Atkinson. Bulletin No. 63, Montana Experiment Station.
- First Annual Report of the Superintendent of Demonstration Farms for North Dakota. North Dakota Experiment Station.
- Campbell's 1907 Soil Culture Manual. H. W. Campbell, Lincoln, Nebraska.
- Proceedings of the First Dry-farming Congress, Denver, Colorado.

- Evaporation Losses in Irrigation and Water Requirements of Crops. S. Fortier. Bulletin 177, Office of Experiment Stations, U. S. Department of Agriculture.
- Cement Pipe for Small Irrigating Systems. G. E. P. Smith. Bulletin 55, Arizona Experiment Station.

1908

- Dry-farming in Idaho. Elias Nelson. Bulletin No. 62, Idaho Experiment Station.
- How to Make Dry-farming Pay: Being the Results of Forty Years of Successful Arid Land Cultivation in Utah. George L. Farrell, Logan, Utah.
- Milling Qualities of Wheat. R. Stewart and J. E. Greaves. Bulletin 103, Utah Experiment Station.
- Milo as a Dry Land Grain Crop. C. R. Ball and A. H. Leidigh. Farmers' Bulletin 322, U. S. Department of Agriculture.
- Dry Land Grains. W. M. Jardine. Circular No. 12, Bureau of Plant Industry, U. S. Department of Agriculture.
- Dry-farm Investigations in Montana. A. Atkinson and J. B. Nelson. Bulletin 74, Montana Experiment Station.
- The Storage of Winter Precipitation in Soils. J. A. Widtsoe, Bulletin No. 104, Utah Experiment Station.

- Dry Land Farming and Kindred Topics. F. S. Cooley. Farmers' Bulletin No. 1, Montana Agricultural College.

- Notes on Dry-farming. W. M. Jardine. Circular No. 10 Bureau of Plant Industry, U. S. Department of Agriculture.

- Second Annual Report of the Superintendent of Demonstration Farms for North Dakota. North Dakota Experiment Station.

- The Plains: Some Press Bulletins. Bulletin 123, Colorado Experiment Station.

- Dry Land Olive Culture in Northern Africa. T. H. Kearney. Bulletin 125, Bureau of Plant Industry, U. S. Department of Agriculture.

- Report of the Second Dry-farming Congress, Salt Lake City, Utah.

- Dry Land Agriculture: Papers Read at the Second Annual Meeting of the Cooperative Experiment Association of the Great Plains Area. Bulletin 130, Bureau of Plant Industry, U. S. Department of Agriculture.

- Report on Dry-farming in America. W. Strawbridge, I. S. O., Adelaide, South Australia.

1909

- The Influence of Depth of Cultivation Upon Soil Bacteria and Their Activities. W. E. King and C. J.

T. Doryland. Bulletin 161, Kansas Experiment Station.

- Tests of Pumping Plants in New Mexico. B. P. Fleming and J. B. Stoneking. Bulletin 73, New Mexico Experiment Station.
- On the Relation of Active Legumes to the Soil Nitrogen of Nebraska Prairies. F. J. Always and R. M. Pickney. The Journal of Industrial and Engineering Chemistry, November, 1909, page 771.
- Tillage and Its Relation to Soil Moisture. C. C. Thom. Popular Bulletin No. 22, Washington Experiment Station.
- Methods of Tillage for Dry-farming. G. Severance. Popular Bulletin No. 15, Washington Experiment Station.
- Dry-farming in Wyoming. V. T. Cooke. State Dry-farming Commission, Cheyenne, Wyoming.
- Dry-farming in Wyoming. J. D. Towar. Bulletin No. 80, Wyoming Experiment Station.
- Report of the Third Dry-farming Congress, Cheyenne, Wyoming.
- Alfalfa in Cultivated Rows for Seed Production in Semiarid Regions. C. J. Brand and J. M. Westgate, Circular No. 24, Bureau of Plant Industry, U. S. Department of Agriculture.

- Factors Influencing Evaporation and Transpiration. J. A. Widtsoe. Bulletin 105, Utah Experiment Station.

- A Study of the Production and Movement of Nitric Nitrogen in an Irrigated Soil. R. Stewart and J. E. Greaves. Bulletin 106, Utah Experiment Station.

- Dry-farming: Report of the Proceedings at the Third Dry-farming Congress and further Investigations in America. Senator J. H. McColl. Government Printing Office, Australia.

- Dry-farming: Its Principles and Practice. W. Macdonald. The Century Company.

1910

- Dry Land Farming in Eastern Colorado. H. M. Cottrell. Bulletin 145, Colorado Experiment Station.

- Dry-farm Practice in Montana. A Atkinson. Circular No. 3, Montana Experiment Station.

- The Fixation of Nitrogen in Some Colorado Soils. W. P. Headeen. Bulletin 155, Colorado Experiment Station.

- Drouth Resistant Plants for the Arid Southwest. J. J. Thornber. Timely Hints of Farmers, No. 83, Arizona Experiment Station.

- The Use of Windmills in Irrigation in the Semi-arid West. P. E. Fuller, Farmers' Bulletin 394, U. S. Department of Agriculture.
- A Contribution to our Knowledge of the Nitrogen Problem under Dry-farming. F. J. Alway and R. S. Trumbell. Journal of Industrial and Engineering Chemistry, April, 1910.
- Report of the Fourth Dry-farming Congress, Billings, Montana.
- Tri-Local Experiments on the Influence of Environment on the Composition of Wheat. J. A. LeClere and S. Leavitt. Bulletin 128, Bureau of Chemistry. U. S. Department of Agriculture.
- Fruit Growing for Home Use in the Central and Southern Great Plains. H. P. Gould. Circular No. 51, Bureau of Plant Industry, U. S. Department of Agriculture.
- Traction Plowing. L. W. Ellis, Bulletin No. 170, Bureau of Plant Industry, U. S. Department of Agriculture.
- Seasonal Nitrification as Influenced by Crops and Tillage. C. A. Jensen. Bulletin No. 173, Bureau of Plant Industry, U. S. Department of Agriculture.
- Cooperative Grain Investigations. Nephi Sub-station. Nephi, Utah. Report of the Years 1907 and 1908 and 1909. F. D. Farrell.
- The Nitrogen and Carbon in the Virgin and Plowed Soils of Eastern Oregon. C. E. Bradley. The

Journal of Industrial and Engineering Chemistry, April, 1910, page 138.

- Storing Moisture in the Soil. W. W. Burr. Bulletin No. 114, Nebraska Experiment Station.

APPENDIX B

THE SMOOT-MONDELL BILL FOR HOME-STEADNIG OF DRY-FARM LANDS

SECTION I

That any person who is a qualified entryman under the homestead laws of the Untied States may enter, by legal subdivisions, under the provisions of this act, in the states of Colorado, Montana, Nevada, Oregon, Utah, Washington and Wyomnig, and the territories of New Mexico and Arizona, 320 acres, or less, of nonmineral nonirrigable, unreserved, and unappropriated surveyed public lands which do not contain merchantable timber, located in a reasonably compact body, and not over one and one half miles in extreme length. Provided, that no lands shall be subject to entry under the provisions of this act until such lands shall have been designated by the secretary of the interior as not being in his opinion susceptible of irrigation at a reasonable cost from any known source of water supply.

SECTION 2

That any person applying to enter land under the provisions of this act shall make and subscribe, before proper officials, an affidavit, as required by Section

2290 of the Revised Statutes, and in addition thereto shall make affidavit that the land sought to be entered is of the character described in Section I of this act and shall pay the fees now required to be paid under the homestead laws.

SECTION 3

That any homestead entryman of lands of the character herein described, upon which final proof has not been made, shall have the right to enter public lands, subject to the provisions of this act, contiguous to his former entry which shall not exceed 320 acres, and residence and cultivation of the original entry shall be deemed as residence upon and cultivation of the additional entry.

SECTION 4

That at the time of making final proof, as provided in Section 2259 of the Revised Statutes, the entryman under this act shall, in addition to the proofs and affidavits required under the said section, prove by two creditable witnesses that at least one eighth of the area embraced in this entry was continuously cultivated to agricultural crops other than prairie grasses, beginning with the second year of the entry, and at least one fourth of the area embraced in the entry was so continuously cultivated beginning with the third year on the entry.

SECTION 5

That nothing herein contained shall be held to affect the right of a qualified entryman to make homestead entry in the states named in Section 2259 of the Revised Statutes, but no person who has made entry under this act shall be entitled to make homestead entry under the provisions of said section, and no entry under this act shall be commuted.

SECTION 6

That whenever the secretary of the interior shall find that any tracts of land, in the state of Utah, subject to entry under this act, do not have upon them such a sufficient supply of water suitable for domestic purposes as would make continuous residence upon the land possible, he may, in his discretion, designate such tracts of land, not to exceed in the aggregate 2,000,000 acres, and thereafter they shall be subject to entry under this act without the necessity of residence. Provided, that in such event the entryman on any such entry shall cultivate in good faith not less than one eighth of the entire area of the entry during the second year, onc fourth during the third year, and one half during fourth and fifth years after the date of such entry, and that after entry, and until final proof, the entryman shall reside within such distance of said land as will enable him successfully to farm the same as required by this section.

